THE ESSENTIAL GUIDE TO
PREALGEBRA

유하림(Harim Yoo) 지음

ESSENTIAL MATH SERIES

Preface

To. 학부모님과 학생들께

압구정 현장 강의를 통해, Prealgebra부터 AP Calculus까지 가르치면서, 강사로서 느낀 점은 학부모님들이 선택할 수 있는 국내에서 판매되는 교재가 몇 없다는 것과 충분히 많은 연구와 노력을 통해, 개발된 교재들은 더더욱 부족하다는 점입니다.

국내 수능 시장에서는 "OOO" 커리큘럼으로 진행되는 수업들과 교재들이 많은데 비해, 미국 SAT 시장과 유학 시장은 그렇지 않았기 때문에, 이렇게 유하림 커리큘럼의 시작을 알리는 것이 기대되면서도 떨립니다. Prealgebra를 처음 배우는 학생들이 흥미롭게 배우길 희망하면서 교재를 썼습니다. 또한, 강의나 교재가 너무 쉽다고 느껴지진 않을 정도의 개념 강의 수준을 유지하기 위해 노력을 했으며, 지금까지 수업해왔던 내용들을 많이 담았기 때문에, 국내에서 유학을 준비하는 학생이나, 해외에서 공부하고 있는 학생들에게 도움이 되길 진심으로 희망해봅니다.

이 교재를 출판할 수 있도록 물심양면 힘써주신 마스터프렙 권주근 대표님께 감사합니다. 또한, 제게 항상 롤모델이 되어주시고, 강사로써 성장할 수 있는 원동력을 주시고 계신 심현성 대표님께 감사의 말씀을 전합니다. 제게 언제나 든든한 지원군이 되는 아내와 딸, 그리고 부모님께도 항상 감사합니다. 마지막으로 제 삶에 이러한 기회를 주신 하나님께 감사합니다. 앞으로 더 좋은 교재를 출판하여 견고하고 튼튼한 유하림 커리큘럼을 완성시키길 희망합니다.

유하림(Harim Yoo)

저자 소개

유하림(Harim Yoo)

미국 Northwestern University,
B.A. in Mathematics and Economics
(노스웨스턴 대학교 수학과/경제학과 졸업)

마스터프렙 수학영역 대표강사
압구정 현장강의 ReachPrep 원장

고등학교 시절 문과였다가, 미국 노스웨스턴 대학교 학부 시절 재학 중 수학에 매료되어, Calculus 및 Multivariable Calculus 조교 활동 및 수학 강의 활동을 해온 문/이과를 아우르는 독특한 이력을 가진 강사이다. 현재 압구정 미국수학/과학전문학원으로 ReachPrep(리치프렙)을 운영 중이며, 미국 명문 보딩스쿨 학생들과 국내 외국인학교 및 국제학교 학생들을 꾸준히 지도하면서 명성을 쌓아가고 있다.

2010년 자기주도학습서인 "몰입공부"를 집필한 이후, 미국 중고교수학에 관심을 본격적으로 가지게 되었고, 현재 유하림커리큘럼 Essential Math Series를 집필하여, 압구정 현장강의 미국수학프리패스를 통해, 압도적으로 많은 학생들의 피드백을 통해, 발전적으로 교재 집필에 힘쓰고 있다.

유학분야 인터넷 강의 1위 사이트인 마스터프렙 수학영역 대표강사 중 한 명으로 미국 수학 커리큘럼의 기초수학부터 경시수학까지 모두 영어와 한국어로 강의하면서, 실전 경험을 쌓아 그 전문성을 확고히 하고 있다.

[저 서] 몰입공부
The Essential Workbook for SAT Math Level 2
Essential Math Series 시리즈

저자직강 인터넷 강의 : 유학 분야 No.1 마스터프렙(www.masterprep.net)

이 책의 특징

유하림 커리큘럼 Essential Math Series의 시작점입니다. 이 책을 처음 접하는 학부모님과 자녀분들을 생각하면서 집필하였습니다. 특히, 미국 명문 Junior Boarding School 및 Boarding School을 진학하고, 성공적으로 적응하기 위해 반드시 필요한 내용이 무엇일까 고민하였고, 6학년(예비7학년)에게 제일 필요한 수학 교재가 무엇일까 생각하면서 일년 내내 작업한 교재입니다.

1st
기본에 충실한 책

이 책을 통해, 혼자서 고민하고 공부할 학생들과 마스터프렙 인강을 통해 공부할 학생들, 그리고 현장강의를 통해 저와 함께 공부할 학생들을 위해, 기본에 가장 충실한 교재를 집필하고자 노력했습니다. 수학의 기초가 있건 없건, 이 교재를 통해 Prealgebra과정에서 반드시 필요한 예제들을 포함시키기 위해 집필하고 개정하는 과정을 여러 번 반복했습니다. 학생의 눈높이에 맞추기 위해, 현장강의 및 인터넷 강의에 대한 학생들의 다양한 반응을 토대로 매번 수정한 노력의 산물입니다.

2nd
생각의 확장을 위한 책

Essential Math Series를 AMC와 같은 경시 수학을 준비하려는 학생을 위한 시리즈로 만들기 위해 예제 선정에 고민하고, 풀이방향을 잡았습니다. 교과 과정의 기본과 더불어 문제 해결의 가장 본질이 되는 개념을 어떻게 설명하고, 어떻게 받아들여야 심화 문제에서 생각을 확장할 수 있을지 고민하며 집필하였습니다. 특히, Prealgebra의 경우, 미국경시수학의 시작인 MATHCOUNTS와 AMC 8에 필요한 생각의 재료를 최대한 담기 위해 노력했습니다.

3rd
유학생을 위한 단 한 권의 책

미국수학을 정말 미국수학답게 가르치기 위해 열심히 공부하고 연구하고, 앞으로도 그러할 것입니다. 노스웨스턴 대학교 학창시절 수학에 대한 열정을 뒤늦게 꽃피워 밤새워 공부했던 것처럼, 저는 학생들을 더 잘 가르치고, 더 나은 미래로 이끌기 위해, AMC, AIME, ARML, HMMT, PUMaC, SUMO와 같은 문제들을 동일한 열정으로 밤낮없이 풀고 해석합니다. 여러분이 지금 보는 이 책은 제 현재 노력의 최선의 산실이며, 앞으로도 그러할 것입니다. 이 책을 통해 수학을 두려워하지 않고, 문제 해결을 즐거워하며, 이른 나이에 수학에 대한 열정을 꽃피우길 기대합니다.

CONTENTS ···

Topic 1
Arithmetic of Numbers

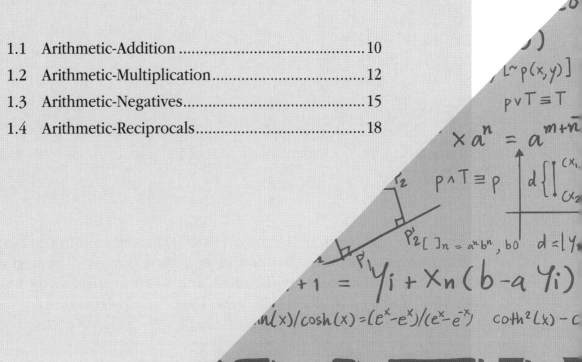

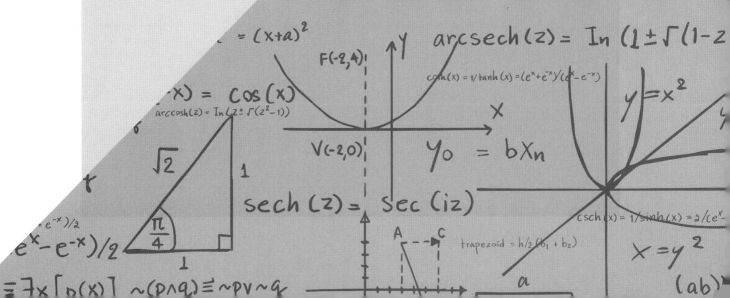

1.1 Arithmetic-Addition

Given three real numbers a, b, and c,

- $a + b = b + a$: commutative property of addition.

- $(a + b) + c = a + (b + c)$: associative property of addition.

Example

Which of the followings are equal to $(5 + 61) + 11 + 19$?
(A) $(61 + 5) + 19 + 11$
(B) $5 + (61 + 11) + 19$
(C) $11 + 5 + 61 + 19$
(D) $(61 + 5 + 11 + 19)$
(E) $11 + (5 + 61 + 19)$

Solution
By the commutative and associative properties of addition, the numbers can be added up in any order and they can be grouped in any way.

---

1 Dan is a farmer who milks cows. His 100 cows line up in the barn and he begins milking them. The first cow gives 2 pints of milk. The next cow gives 3 pints of milk. The next gives 4 pints, and so on, increasing by 1 pint each cow. How many pints of milk does the 100th cow produce?

Problems about number addition always deal with the order of addition. If there are many numbers to add, think about a pair of numbers that can be easily computed, instead of adding from left to right.

Example

Compute $123 + 142 + 158 + 177$.

Solution
The quickest way to figure out is that $123 + 177 = 300$ and $142 + 158 = 300$. Hence, the sum of four numbers is 600.

2 What is the sum of the first 40 positive odd integers?

1.2 Arithmetic-Multiplication

Given three numbers a, b, and c

- $ab = ba$: commutative property of multiplication.

- $(ab)c = a(bc)$: associative property of multiplication.

- $(a+b)c = ac+bc$: distributive property of numbers.

Example

Find $2 \cdot 3 \cdot 25 \cdot 5 \cdot 4 \cdot 10$.

Solution

Let's use the <u>commutative</u> and <u>associative</u> properties of multiplication to simplify these numbers. We notice that $2 \cdot 5$ and $25 \cdot 4$ will be easy to multiply together, so we can use the commutative property to change the order of multiplication:

$$2 \cdot 3 \cdot 25 \cdot 5 \cdot 4 \cdot 10 = 2 \cdot 5 \cdot 25 \cdot 4 \cdot 3 \cdot 10.$$

Then we can use the associative property to group these numbers:

$$10 \cdot 100 \cdot 3 \cdot 10 = \boxed{30,000}.$$

3 What is the product of all non-negative one-digit integers?

4 Ten pictures, each measuring 4 inches by 6 inches, are placed on a piece of green poster board so that there is no overlap, the size of which measures 20 inches by 17 inches. Assume that the shapes of all pictures are rectangular. How many square inches of the green background will show after the pictures are placed?

5 Compute $115 \times 205 + 35 \times 205$ in your head. (Hint: Try use the distributive property of real numbers.)

6 Compute the following expressions.

(a) $117 \cdot 2001 + 9 \cdot 1999 \cdot 13$ (b) $234 \cdot 97 + 2 \cdot 9 \cdot 13 \cdot 103$

7 Compute $12 \cdot 45 + 12 \cdot 32 + 12 \cdot 23$ by using the distributive property of real numbers.

1.3 Arithmetic-Negatives

Given a real number x, then there is a number $-x$ such that $x + (-x) = 0$. In other words, $-(-x) = x$.

Example

Calculate:
$$4(2)(-3) + (-(-2))(3).$$

Solution

We start by looking at the first term: $4(2)(-3) = 8(-3) = -8 \cdot 3 = -24$. Then we look at the second term: when we have two negatives being multiplied, they cancel to create a positive, so $(-(-2)) = 2$ and $(-(-2))(3) = 2(3) = 6$. Thus, our desired sum calculates to

$$-24 + 6 = -(24 - 6) = \boxed{-18}.$$

Algebraic properties we use about the negatives are given such that

- Even number of negatives is positive.

- Odd number of negatives is negative.

8 Find the value of $-(-(-(-3))) + (-2) \cdot (-(-(-3)))$.

9 Compute the following expressions.

(a) $-[9 \cdot (-6) + (-4) \cdot (-5)] + 3 \cdot (-7)$

(b) $[(-3) \cdot (-2) + (-1) \cdot 3] - 2 \cdot (-3)$

10 Adam and Bob are in a classroom with a chalkboard. They are standing on different halves of the board, and on each half, the number 2 is written. When Adam's teacher gives a signal, Adam multiplies the number on his side of the board by -1 and writes the answer on the board, erasing the number he started with. Bob does the same on each signal, except that he multiplies by 1. The teacher gives 50 signals in total. How many times (including the initial number) do Adam and Bob have the same number written on the board (including those 2's at the beginning)?

11 Compute the following expressions.

$$(-3) \cdot (-(-(-(-2)))) \cdot (-(-2)) \cdot (-(-(-(-(-3)))))$$

12 Ten boys are sitting in a row of chairs, each of whom is thinking of a negative integer greater than or equal to -15. Each boy subtracts, from his own number, the number of the adjacent boy sitting to his right. Bob is sitting at the rightmost position. Because he has nothing else to do, Bob observes all the differences and notices that all of the values are positive. Let x be the greatest integer owned by any of the 10 people at the beginning. What is the minimum possible value of x?

1.4 Arithmetic-Reciprocals

Given a nonzero a, then the reciprocal of a is $\dfrac{1}{a}$ such that $\dfrac{1}{a} \times a = 1$.

Example

Find the reciprocal of the sum of the reciprocals of $\dfrac{1}{-5}$ and $-\dfrac{1}{6}$.

Solution

The reciprocal of the reciprocal of a nonzero number is the original number, or $\dfrac{1}{1/x} = x$, so the reciprocal of $\dfrac{1}{-5}$ is -5. Also recall that the negation of a reciprocal is the reciprocal of the negation, or $-\dfrac{1}{x} = \dfrac{1}{-x}$, so we can rewrite $-\dfrac{1}{6}$ as $\dfrac{1}{-6}$. Thus, the reciprocal of $-\dfrac{1}{6}$ is -6. Adding these, we get $(-5) + (-6) = -11$. The reciprocal of -11 is

$$\frac{1}{-11} = \boxed{-\frac{1}{11}}.$$

[13] Does 0 have a reciprocal? Hence, if x is a nonzero real number, compute

$$x \cdot \left(-\frac{1}{x}\right) - (-x) \cdot \left(\frac{1}{x}\right)$$

We can evaluate a mathematical expression by substituting given numerical values to the variables (in other words, placeholders). Let's have a look at how we evaluate a mathematical expression by solving the following example.

Example

Evaluate $\dfrac{1}{-a} + \dfrac{3}{b}$ if $a = -\dfrac{1}{2}$ and $b = \dfrac{1}{3}$.

Solution

Let's remind ourselves of the meaning of $\dfrac{1}{a}$, which is a reciprocal of a. Since the reciprocal of $-\dfrac{1}{2}$ is -2 and there is another negative sign in $-\dfrac{1}{a}$, we get positive 2. Similarly, $\dfrac{3}{b}$ turns into $3 \times \dfrac{1}{b}$, which is equal to $3 \times 3 = 9$. Hence, $\dfrac{1}{-a} + \dfrac{3}{b} = 2 + 9 = 11$.

14 Given $(a, b, c, d) = \left(\dfrac{1}{3}, \dfrac{1}{4}, 2, -\dfrac{1}{8} \right)$, i.e., $a = \dfrac{1}{3}$, $b = \dfrac{1}{4}$, $c = 2$, and $d = -\dfrac{1}{8}$, evaluate

$$\frac{1}{(-a) \cdot (-1/c)} + \frac{1}{b \cdot d}$$

1 Evaluate the following expressions.

(a) $67 + 5 + 3 + 95$

(b) $290 + 9 + 2 + 492 + 5 + 393 + 3 + 0 + 1 + 5$

(c) $120 + 9 - 3 + 4(2 + 3) - 110$

2 Bob was collecting tickets for a school orchestra, trying to determine how many people would attend the show. One group of students gave him 14 tickets, another group 35 tickets, and a third group 23 tickets. Later on, three more groups arrived. A fourth group gave him 15 tickets, a fifth group 16 tickets, and a sixth group 22 tickets. Finally, Bob collected 102 individual tickets. How many tickets in total did Bob collect?

3 Gauss, one of the most renowned mathematicians, used the commutative property so as to find out the sum of the natural numbers from 1 to 100. Mimicking what he did when he was young, find the sum of positive integers from 21 to 29, inclusive.

4 Compute $(12131 + 22136) \cdot 4 \cdot 0 \cdot 2 + (1234 \cdot 1 + (-1233)) \cdot 173$.

5 Bob has $25 in his bank account. His father promises to triple the amount in his bank account. Secretly, Bob's mother also promises to double the amount Bob has in his bank account. Finally, Bob's uncle says he will quadruple(=multiply by four) whichever amount Bob has in the bank account at the moment. Bob first asks his dad to triple the bank account balance, his mother to double his current balance, and finally his uncle to quadruple what he has. How many dollars does he have in his bank account right after he puts his uncle's money to it?

6 Bob owns a tangerine tree. The tree has 10 main branches, each branch of which has exactly 2 smaller branches. Furthermore, each smaller branch has 3 tangerines, always with 2 leaves surrounding each tangerine. How many leaves are there on his tree?

7 Evaluate $-(-(-(-(-(-(-3))))))\cdot(-(-(-4)))\cdot(-(-(-(-(-(-5)))))).$

8 Suppose there is a square with side lengths of 2 units. A second square is formed by having sides that are **150%** of the length of the sides of the first square. Similarly, a third square is formed by having sides that are **150%** of the length of the sides of the second square. What is the difference between the area of third square and that of the original square?

9 Let $a = b = \dfrac{1}{2}$, $c = 3$, and $d = -\dfrac{1}{4}$. Calculate

$$\frac{1}{a \cdot \dfrac{1}{c}} - \frac{b}{a \cdot d}$$

10 Two congruent(=equal in shape and size) squares are placed side by side. The perimeter of the rectangle(=a newly formed figure) is 60 units. What is the area of the rectangle, in square units? (Recall that the area of rectangle is the product of base length and height.)

11 Since the extreme drought that started in 2010, Korean farmers have received $12 million in water subsidies(=financial support). Specifically, Korean cotton and rice farmers received an additional $130 million. What is the maximum amount of water subsidies Korean farmers, including the cotton and rice farmers, could have received in total?

1

(a) 170   (b) 1200   (c) 36

2 227 tickets in total.

3 225

4 173

5 600 dollars.

6 120 leaves.

7 60

8 16.25

9 10

10 200 square inches.

11 142 million dollars.

Topic 2

Raising Powers

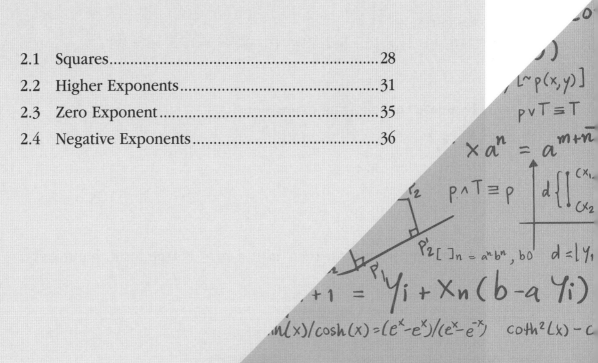

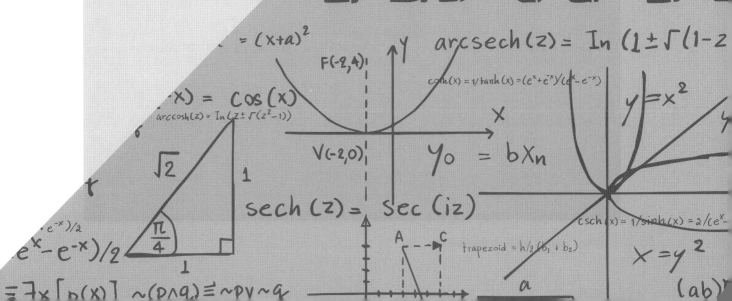

2.1 Squares

We call the product of a number and itself a square. We can write a square as a power using 2 as the exponent. For example, $3^2 = 3 \cdot 3$. When speaking, we say "3 squared" for 3^2, and "squaring" a number means multiplying the number by itself.

> ### Example
>
> Simplify $16 - 4 \cdot 2^2$.
>
> #### Solution
> $16 - 4 \cdot 2^2 = 16 - 4(2 \cdot 2) = 16 - 4(4) = 16 - 16 = 0$.

1 Evaluate the following expressions.

(a) $(4 \cdot 5)^2$

(b) $4^2 \cdot 5^2$

(c) Hence, determine whether $(ab)^2 = a^2 b^2$ is true for a, b numbers.

2 Evaluate the following expressions.

(a) $(512 \div 16)^2$

(b) $512^2 \div 16^2$

(c) Hence, determine whether $\left(\dfrac{a}{b}\right)^2 = \dfrac{a^2}{b^2}$ is true for a, $b(\neq 0)$ numbers.

When we perform arithmetic, we use the rule called PEMDAS, which stands for

Parentheses → Exponents → Multiplication
→ Division → Addition → Subtraction

3 Evaluate the following expressions using PEMDAS.

(a) $8 + 6(3 - 8)^2$

(c) $92 - 45 \div (3 \cdot 5) - 5^2$

(b) $5(3 + 4 \cdot 2) - 6^2$

(d) $8\left(6^2 - 3(11)\right) \div 8 + 3$

A solid definition of perfect squares can be laid out as in the following color box.

An integer n is a perfect square if there is an integer p such that $n = p^2$ and $|p| \le n$.

For instance, 0 is a perfect square because $0 = 0^2$. Likewise, 1 is a perfect square because $1 = 1^2 = (-1)^2$.

On the other hand, 3 is NOT a perfect square because we cannot find any integer such that the product of itself does not produce 3. Let's look at the following examples to figure out the list of perfect squares smaller than 100.

Example

List all positive perfect squares smaller than or equal to 100.

Solution
There are 10 perfect squares, since $10^2 = 100$. In fact, we can count them directly by squaring integers, i.e., $1^2 = 1$, $2^2 = 4$, $\cdots$, $10^2 = 100$. Hence, the list contains $1, 4, 9, 16, 25, 36, 49, 64, 81, 100$.

4 How many perfect squares are between 1000 and 2000?

5 Suppose we know $1^2 + 2^2 + 3^2 + 4^2 + \cdots + 25^2 = 5525$. Use some algebraic properties to evaluate the following expression.

$$3^2 + 6^2 + 9^2 + 12^2 + \cdots + 75^2$$

2.2 Higher Exponents

Let a be any number and let n be a positive integer. The power a^n, pronounced "a to the n," is defined by the equation

$$a^n = \underbrace{a \times a \times \cdots \times a}_{n \text{ copies of } a}.$$

For example, $a^5 = a \cdot a \cdot a \cdot a \cdot a$.

We can generalize the algebraic properties of higher exponents as we checked in the squares. Let a and b be numbers and n be a positive integer. Then, the following properties hold.

- Power of product: $(ab)^n = a^n b^n$.

- Power of reciprocal: If b is nonzero, then $\left(\dfrac{1}{b}\right)^n = \dfrac{1}{b^n}$.

- Power of quotient: If b is nonzero, then $(a \div b)^n = a^n \div b^n$.

Example

Simplify 2^{2^2}.

Solution
$2^{2^2} = 2^4 = 16$. Normally, we compute the power expressions in the exponents first.

$\boxed{6}$ Evaluate $(-1)^{(4^2)} + 1^{(2^4)}$.

7 Express the following expression as a power of 3.

$$3^{200} + 3^{200} + 3^{200}$$

8 Evaluate the following expressions.

(a) $\left(2^4\right)^3$

(b) $2^{4 \cdot 3}$

(c) Hence, determine whether $(a^m)^n = a^{mn}$ is true for positive real number a and positive integers m and n.

Suppose we wish to COMPARE two exponential expressions with large exponents. Let's have a look at the following example to investigate what we should do.

Compare 2^{200} and 3^{100}.

Solution
Instead of computing two numbers, we can compare the bases. For instance, $2^{200} = 4^{100}$ and 3^{100} can be compared when we look at bases. Since $4 > 3$, $2^{200} > 3^{100}$.

The idea of solving this type of inequality is to use the two strategies,i.e.,

- Compare the bases by setting the exponents equal.

- Compare the exponents by setting the bases equal.

9 Rearrange 11^{30000}, 5^{40000}, and 2^{80000} in increasing order.

10 Evaluate the value of the following sum of powers of 1.

$$1^2 + 1^4 + 1^6 + 1^8 + \cdots + 1^{200}$$

11 Evaluate the following expressions.

(a) $2^4 + 2^4 + 2^4 + 2^4$

(b) $(2^5 + 2^6 + 2^7) \div 2^3$

(c) $3^5(2^3) - 2^4(3^4)$

(d) $88{,}888^4 \div 22{,}222^4$

(e) $4^4 + 4(4^4) + 6(4^4) + 4(4^4) + 4^4$

(f) $4^{3^3} \div (4^3)^3$ (as a power of 2)

2.3 Zero Exponent

Let m and n be positive integers such that m is greater than n. In the last section, we introduced the quotient of powers (same base) rule:

$$2^{m-n} = 2^m \div 2^n.$$

Let a be any number. Then a^0 is defined to be 1. This definition includes $0^0 = 1$. We will not encounter 0^0 often, and rigorous proof will be laid out in the course called Calculus. However, we define it as 1 in order to satisfy $a^0 \cdot a^n = a^n$ for all valid base a.

$\boxed{12}$ Let $P = (1 - 2 - 3 + 4)^{123456789}$ and $Q = (-1 + 2 + 3 - 4)^{123456789}$. What is the value of

$$(1 + 2 + 3 + 4)^{P+Q}?$$

$\boxed{13}$ Let x be a number. Simplify $6^0 \cdot x^{9876} + 6 \cdot x^{9876}$. Express your answer as a number times a power of x.

2.4 Negative Exponents

Let a be a nonzero number. Let n be a positive integer. Then a^{-n} is defined to be the reciprocal of a^n, so

$$a^{-n} = \frac{1}{a^n}$$

- Let a be a nonzero number and m and n be integers. Then, we have

$$a^m a^n = a^{m+n}$$

- Let a be nonzero and let n be a positive integer. Then, we have

$$\frac{1}{a^{-n}} = a^n$$

- Let a be nonzero and n be a positive integer. Then, we have

$$\left(\frac{1}{a}\right)^{-n} = a^n$$

14 Evaluate the following expressions.

(a) 1^{-5}

(b) 10^{-4}

(c) 2^{-3}

(d) $56 \cdot 2^{-3}$

(e) $56 \div 2^{-3}$

(f) $\left(\frac{1}{16}\right)^{-(-2)}$

15 Compute each of the following expressions.

(a) $3^{-1} \cdot 3^{-2}$

(b) $3^{15} \cdot 3^{-5} \cdot 3^{-4} \cdot 3^{-3}$

16 Evaluate the following expressions.

(a) $\dfrac{1}{2^{-3}}$

(b) $\dfrac{1}{5^{-2}}$

(c) Hence, if a is nonzero real number and n be a positive integer, rewrite $\dfrac{1}{a^{-n}}$ in powers of a.

1 Evaluate the following expressions.

(a) $4 - 8\left((-2)^2 - 4(-3)\right)$

(b) $(5-2)^2 + (2-5)^3$

(c) $5 \cdot 2^5 - (2 \cdot 3)^2$

(d) $5 + (-6)^3 \div (2 \cdot 3^2)$

2 It is convenient for us to remember powers of 2. Let's have a look at a few important powers of 2.

- $2^8 = 256$

- $2^9 = 512$

- $2^{10} = 1024$

Hence, by how much does 4^5 exceed 5^4?

3 Having a closed set of parentheses sometimes makes a difference, especially when there is a negative number. For instance, $-2^2 \neq (-2)^2$. The left-side of the equation above is -4, whereas the right-side is 4. Hence, what is the value of the sum

$$-1^{1234} + (-1)^{2345} + 1^{3456} - 1^{4567}?$$

4 For what positive integer n is $n^2 = 2^6$?

5 How many positive perfect squares are less than $40,000$?

6 Let n be a positive integer. If

$$(1+2+3+4+5+6)^2 = 1^3 + 2^3 + \cdots + n^3,$$

what is the value of n?

7 If n is an integer such that $n^3 = -4913$, what is the value of n?

8 Given an expression

$$b^x$$

we call b as the base and x as the exponent. Here, b is a positive number, normally not equal to 1. Hence, evaluate $5 \cdot x^y - 6 \cdot y^x$ when $x = 3$ and $y = 4$.

1

(a) −124

(b) −18

(c) 124

(d) −7

2 399

3 −2

4 8

5 199 perfect squares

6 $n = 6$

7 −17

8 21

Topic 3

Integers

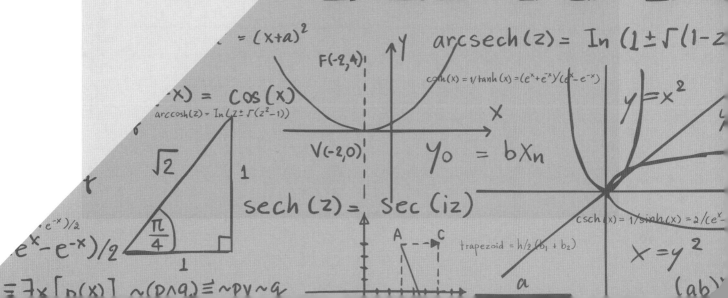

3.1 Multiples

Let a and b be numbers, especially integers. We say that a is a multiple of b if a equals b times some integer. In other words, a is a multiple of b if there is an integer n such that

$$a = bn$$

For instance, 5 is not a multiple of 2, for if 5 is divided by 2, then the quotient is 2 and the remainder is 1. In this chapter, we will focus on finding the quotient and remainder as well.

Example

We know that 6 and 9 are multiples of 3. Then, is $15 = 6 + 9$ a multiple of 3?

Solution
Yes, we can rewrite $6 + 9 = 2(3) + 3(3) = (2+3)(3) = 5(3)$, which is a multiple of 3.

$\boxed{1}$ Notice that both 48 and 144 are multiples of 12. What can you make of 192? Is $192(= 48 + 144)$ a multiple of 12 as well?

$\boxed{2}$ Suppose x is a multiple of 5. In other words, $x = 5k$ for some integer k. Is $x - 10$ a multiple of 5?

(a) If x is a positive multiple of 2, then is $x+3$ a multiple of 3 for all x?

Solution
No. Since $x = 2(n)$ for some natural number n, $x+3 = 2n+3$ is not divisible by 3 if $n = 1$.

(b) If x is a positive multiple of 2, then is $x+3$ a multiple of 3 for some x?

Solution
Yes. Since $x = 2(n)$ for some natural number n, $x+3 = 2n+3$ can be divisible by 3 if $n = 3$.

(c) If x is a multiple of 2, then is it possible for $x+3$ to be a multiple of 3 for some x?

Solution
Yes. Since $x = 2(n)$ for some natural number n, $x+3 = 2n+3$ is divisible by 3 if $n = 0$.

3 Suppose x is a multiple of 3. Is it possible for $x+13$ to be a multiple of 3?

4 Which integer between 300 and 400, not inclusive, is both a perfect cube and a multiple of 7?

How many odd integers are there from 1 to 100, inclusive?

Solution

Out of 100 numbers, if we take out even numbers, we get all the odd numbers. Since $100/2 = 50$ even numbers can be counted, we get $50(= 100 - 50)$ odd numbers.

On the other hand, we can count them by making a 1-to-1 correspondence. Let x_n be the odd number at the nth position. Then, $x_1 = 1$, $x_2 = 3$, $x_3 = 5$, and so on. We can find the rule of $x_n = 2n - 1$ by inspection. Hence, $x_n = 2n - 1 < 100$ implies that the maximum value of n must be $n = 50$. Therefore, there are 50 odd numbers.

5 Find the largest three-digit multiple of 17.

6 How many integers between 3 and 1001, inclusive, are multiples of 7? (If it is not inclusive, then the answer may differ by 1.)

3.2 Divisibility Test

Let a be an integer and b be a nonzero integer. We say that a is divisible by b if $a \div b$ is an integer.

Here is the summary of divisibility by numbers.

- 2 : Units digit of n is $0, 2, 4, 6$ or 8.

- 3 : Sum of digits of n is a multiple of 3.

- 4 : Number formed by last two digits of n is a multiple of 4.

- 5 : Units digit of n is 0 or 5.

- 8 : Number formed by last three digits of n is a multiple of 8.

- 9 : Sum of the digits of n is a multiple of 9.

- 10 : Units digit of n is 0.

Example

Justify that a multiple of 3 has the sum of digits a multiple of 3 with any example.(The case study of a multiple of 9 is exactly same as the one laid out in the following proof.)

Solution

We will look at a positive integer 267. This is a multiple of 3 because $3 \times 89 = 267$. Is there an easy way to figure out whether it is a multiple of 3?

Notice that $267 = 2(100) + 6(10) + 7$. Then, $2(99 + 1) + 6(9 + 1) + 7 = 2(99) + 6(9) + (2 + 6 + 7)$. Notice that the first two terms are multiples of 3. Therefore, $3(66 + 18) + (2 + 6 + 7)$. If the last expression $(2 + 6 + 7)$ is divisible by 3, then we can write $267 = 3 \times$ some expressions. Therefore, if N is a multiple of 3, then the sum of N's digits must be divisible by 3.

Knowing that multiples of 3 have the sum of digits divisible by 3 is helpful to figure out the prime factorization of that number.

7 It is easy to check that 3, 6, and 9 are nonzero multiples of 3. Any summation of these three numbers always results in a multiple of 3. Hence, determine whether 363637 is divisible by 3.

8 We conventionally write a four-digit number $\overline{ABCD}$ to indicate $A \cdot 10^3 + B \cdot 10^2 + C \cdot 10 + D$, where $1 \leq A \leq 9$ and $0 \leq B, C, D, \leq 9$. If X is the units digit in the four-digit number $\overline{463X}$, divisible by both 3 and 4, then what is the sum of all possible values of X? (Hint : There are two possible values of X, but there is only one value of X satisfying the given condition.)

9 If Y is the tens digit in the five-digit number $\overline{246Y8}$, which is divisible by both 3 and 4, what is the only possible value of Y, where $0 \leq Y \leq 9$?

10 What is the only possible digit Z for which the number $\overline{214Z07}$ is divisible by 9?

3.3 Primes and Composites

A prime number (or simply prime) is a natural number p greater than 1[1] whose only positive divisors are 1 and p. There are three major types of prime numbers we can use.

- 2 : the only even prime.

- 3 : the smallest odd prime.

- $6k \pm 1$: the other primes that are adjacent to multiples of 6.

On the other hand, a composite number (or simply composite) is a natural number c with some positive divisor besides 1 and c itself. In fact, a composite number c can be written as the product of two natural numbers that are both between 1 and itself:

$$c = ab,$$

where a and b are (not necessarily distinct) divisors of the composite number c.

Example

List the first 10 primes.

Solution
First, 2 and 3 are given. Now, let's find out the other primes using the fact that they are adjacent to multiples of 6.

Case 1) 5, 7, both of which are adjacent to 6.

Case 2) 11, 13, both of which are adjacent to 12.

Case 3) 17, 19, both of which are adjacent to 18.

Case 4) 23, 25, one of which is prime. (25 is not prime.)

Case 5) 29, 31, both of which are adjacent to 30.

As one can notice from case 4, if not all adjacent numbers to multiples of 6 are prime, then one of the two numbers MUST be prime. For example, for 24, 25 is not prime, but 23 is prime.

---

[1] 1 is the only natural number that is neither prime nor composite. It has only one positive divisor, i.e., itself. This is one reason why 1 is a very special number. It is called the unity.

11 List down the prime numbers greater than 90 and less than 100.

12 Bob teaches a fourth grade class at an elementary school whose class sizes are always at least 30 students and at most 36. One day Bob decides that he wants to arrange the students who sat on their individual desks in a rectangular grid with no gaps. Unfortunately for Bob, he discovers that doing so could only result in one straight row(or column) of desks. How many students does Bob have in his class?

```
     2  3  4  5  6  7  8  9  10
11  12 13 14 15 16 17 18 19 20
21  22 23 24 25 26 27 28 29 30
31  32 33 34 35 36 37 38 39 40
41  42 43 44 45 46 47 48 49 50
 ⋮  ⋮  ⋮  ⋮  ⋮  ⋮  ⋮  ⋮  ⋮  ⋮
```

Follow the instructions and the grid above will reveal information about the primes.

- Circle the number 2.

- Cross out all of the multiples of 2 that are greater than 2.

- Circle the smallest number that is neither circled nor crossed out.

- Cross out all of the multiples of the integer you circled in the third bullet point, which have not already been crossed out.

- Repeat the third and fourth bullet points until all of the numbers are either circled or crossed out.

We can perform this process with larger lists of integers starting from 2. In this way we identify which integers in the list are prime or composite. This process is known as the Sieve[2] of Eratosthenes after the Greek mathematician Eratosthenes who devised the algorithm in roughly around 230 BC.

[13] What is the largest prime number less than 200, none of whose digits is composite?

---

[2]A sieve is a utensil of wire mesh in order to distinguish large matters from small matters. Smaller material passes through the wire mesh while larger material is stopped by the sieve.

14 Find

(a) the smallest prime divisor of **9**

(d) the smallest prime divisor of **25**

(b) the smallest prime divisor of **15**

(e) the smallest prime divosor of **35**

(c) the smallest prime divisor of **16**

(f) the smallest prime divisor of **49**.

When checking to see if a natural number is prime or composite, we only need to test for divisibility by all the primes up to the square root of that natural number. If one of those primes is a divisor, the natural number is composite. Otherwise, it is prime. That's much easier than performing the whole Sieve of Eratosthenes up to that natural number! This quicker method will save us much more time as number theory problems turn more difficult.

15 Identify which of the following integers are prime.

(a) 197

(d) 499

(b) 297

(e) 553

(c) 323

(f) 773

16 We say a natural number n can be prime factorized if $n = p_1^{n_1} p_2^{n_2} \cdots p_m^{n_m}$ where $p_1, p_2, \cdots p_m$ are distinct prime numbers. In any case, n can be written as a unique product of primes. Hence, how many prime numbers are multiples of 3?

17 Determine which of the following natural numbers are prime and which are composite.

(a) 313

(b) 391

(c) 569

(d) 853

The prime factorization of an integer uses power expressions. For example,

$$12 = 2^2 \cdot 3^1$$

Normally, all the primes are written in increasing order. Also, we do not write 1 in the exponent, so we leave it unwritten.

In Prealgebra, we learn how to make a factor tree, some of which you might have learned in previous years of mathematics. Look at the following branch of factor trees of 12.

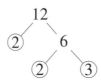

As one can see from the nodes, we circle primes in the tree, so we circle the 2 as shown. Then, we make a progression of branching out to the numbers that are not circled, i.e., 6.

We can factor 6 into the product of two smaller numbers, 2 and 3. Each of these are prime, so we circle both. All numbers are circled, so we are done. Hence, we get the prime factorization of 12 from the factor tree.

$$12 = 2 \cdot 2 \cdot 3 = 2^2 \cdot 3^1$$

Example

Let's prime factorize 288 using the factor tree.

Solution

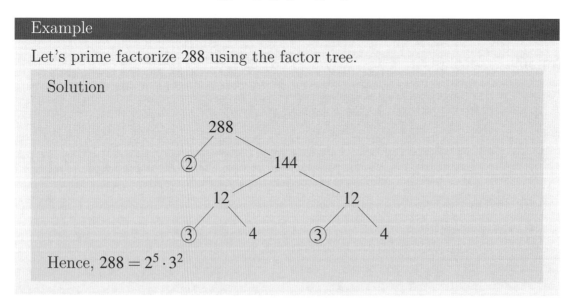

Hence, $288 = 2^5 \cdot 3^2$

In fact, any positive integer can be factorized into the product of primes. Natural numbers can be categorized into three possible number types, i.e., the primes, the composites, and 1.

- The prime factorization of a prime number is simply the number itself. For example, the prime factorization of 5 is 5.

- The following process is due to the Well-Ordering Principle of natural numbers. The WOP can be explained in the following illustration of prime factorization.

  Given any composite number, it can be decomposed into the product of two smaller numbers, by its definition. We continue making more sub-branches with these smaller positive numbers. In each case, we either identify the number as prime to circle it, or we break the number into a product of smaller positive integers.

  What is the smallest natural number? It's 1, right? As we decompose the given number into the product of two smaller numbers, the numbers written on the nodes approach 1. In other words, the whole process must end with a finite number of steps because we can't keep producing smaller and smaller positive integers. So, there must be a certain point at which we can't continue the factor tree. At this point, all of the numbers at the end nodes of the tree must be prime since they cannot be factored anymore. Now, we are able to find a prime factorization of any positive integer.

- Here is the definition of the prime factorization of 1, i.e.,

$$1 = 1$$

  Simple. Right? This isn't a product of primes, but this is necessary when we talk about prime factorization of all positive integers.

The prime factorization of each number is unique. That is, for any positive integer besides 1, there is only one group of primes whose product equals the integer. (Here, we disregard the order of multiplication.)

Since there is a mention of the Well-Ordering Principle (a.k.a. WOP), let's briefly demonstrate what this principle is.

> Given any nonempty set X of natural numbers, there exists the minimum element of X. For instance, if $X = \{3, 1, 4\}$, then the minimum element of X is 1.

Normally, we write the prime factorization of $n = p_1^{n_1} p_2^{n_2} \cdots p_m^{n_m}$ where $p_1, p_2,$ $\cdots, p_m$ are distinct prime numbers, making sure that $p_1 < p_2 < p_3 < \cdots < p_m$.

18 Find the prime factorization of each of the following numbers.

(a) 30

(b) 60

(c) 252

(d) 243

3.5 Least Common Multiple

Given two positive integers p and q, the least common multiple l, denoted by $\text{lcm}[p,q]$, is the least integer(positive, of course) that is a multiple of both p and q. Usually, we take the maximum exponents of both p and q, completely factorized.

- $\text{lcm}[12,15] = 60$ because $12 = 2^2 \cdot 3, 15 = 3 \cdot 5$ so $\text{lcm}[12,15] = 2^2 \cdot 3^1 \cdot 5^1 = 60$.

- $\text{lcm}[5,6] = 30$. If two numbers are co-prime[3], then the least common multiple is the product of the two numbers.

Example

Iphone rings its bells every 12 minutes, Samsung Galaxy rings its bells every 20 minutes, and LG G-series rings its bells every 25 minutes. If they all ring their bells at noon on the same day, at what time will they next all ring their bells together?

Solution
What we compute is the least common multiple of 12, 20 and 25.

$$12 = 2^2 \cdot 3^1$$
$$20 = 2^2 \cdot 5^1$$
$$25 = 5^2$$

We choose the maximum exponents of each prime that appears in the list at least once. Therefore, the least common multiple is $\text{lcm}[12,20,25] = 2^2 \cdot 3^1 \cdot 5^2 = 300$(minutes). In hours, 300 minutes equal 5 hours. Therefore, all of the phones will ring together at $05:00$ P.M. the same day.

19 Compute each of the following least common multiples.

(a) $\text{lcm}[96,144]$ (b) $\text{lcm}[24,36,42]$

---

[3]We say two numbers a and b are coprime(or co-prime) if the only common divisor between the two is 1.

3.6 Greatest Common Divisor

Given two integers p and q, the greatest common divisor d, denoted by $\gcd(p,q)$, is the largest integer that divides both p and q. Usually, we take the minimum exponents of both p and q, completely factorized.

- The greatest common divisor between 12 and 15 is 3 because $12 = 2^2 \cdot 3$ and $15 = 3 \cdot 5$.

- The greatest common divisor between 5 and 6 is 1. In this case, we say 5 and 6 are co-prime.

Example

Compute $\gcd(27,45)$.

Solution
Let's prime factorize 27 and 45.

$$27 = 3^3$$
$$45 = 3^2 \cdot 5^1$$

We choose the minimum exponents of each prime that appears in the list more than once. Therefore, the greatest common divisor is $\gcd(27,45) = 3^2 = 9$.

20 Compute the following greatest common divisors.

(a) $\gcd(72,240,288)$ \qquad\qquad (b) $\gcd(14,42)$

1 Find the prime factorization of the following numbers.

(a) 3072

(b) 1236

2 Find the remainder when 12345678×27 divided by 9?

3 What is the sum of the three smallest multiples of 7 that are greater than 100?

4 Is 987654321 divisible by 9? If yes, justify your answer.

5 Compute the following expressions.

(a) lcm[40, 30, 65]

(b) gcd(248, 364)

6 What is the units digit of the product of any consecutive five positive integers?

7 Find the smallest prime factor of $23^{19} + 41^{13}$.

1

(a) $2^{10} \cdot 3^1$ (b) $2^2 \cdot 3^1 \cdot 103^1$

2 0

3 336

4 987654321 is divisible by 9 because the sum of digits is a multiple of 9.

5

(a) $\operatorname{lcm}[40, 30, 65] = 1560$ (b) $\gcd(248, 364) = 4$

6 0

7 2

Topic 4
Fractions

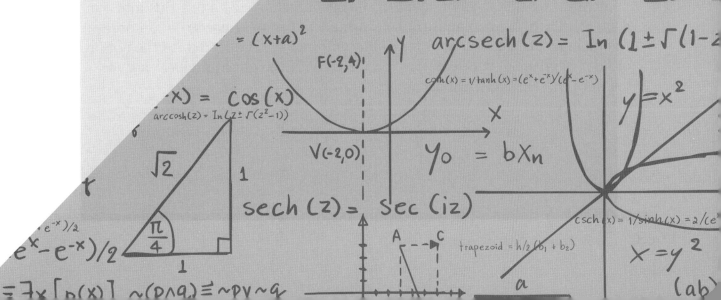

4.1 Fraction

The set of integers is extended in order to solve mathematical problems with divisions. Let's define the set of rational numbers, or fractions, in the following manner.

Given any a and $b \in \mathbb{Z}$ where $b \neq 0$, then

$$\mathbb{Q} = \{\frac{a}{b} \mid \gcd(a,b) = 1\}.$$

A fraction such as $\frac{a}{b}$ has two parts - a known as numerator and b as denominator.

- If x is not zero, then $\frac{0}{x} = 0$.

- If x is not zero, then $\frac{x}{x} = 1$.

- $\frac{x}{1} = x$.

- If y is not zero, then $\frac{-x}{y} = -\frac{x}{y}$.

- If y is not zero, then $\frac{x}{-y} = -\frac{x}{y}$.

- If y is not zero, then $\frac{-x}{-y} = \frac{x}{y}$.

Now, let's have a look at what a fraction teaches us, and how we use this knowledge into problem-solving skill set. Let's suppose we have 17/7. Then, $17/7 = 2 + 3/7$. Here, 2 is called the quotient and 3 the remainder.

Conventionally, a positive fraction tells us two pieces of information - quotient and remainder. Here, we will focus on the quotient. The quotient, in a positive fraction, tells us the number of positive multiples of the numerator, smaller than or equal to the denominator. Let's put this into our context. There are two multiples of 7, smaller than or equal to 17. In order to eliminate the not-so-useful remainder information, we introduce the notion of floor expression, which simply rounds DOWN the given number. For instance, in our numerical example,

$$\lfloor \frac{17}{7} \rfloor = \lfloor 2 + \frac{3}{7} \rfloor = 2$$

For how many values of n between 1 and 100 is $\frac{n}{2}$ an integer?

Solution

We use the floor expression to find out the number of multiples of 2.

$$\lfloor \frac{100}{2} \rfloor = 50$$

There are 50 multiples of 2 between 1 and 100, inclusive. Let's investigate more about the floor notations. Floor notations get rid of all the floating parts (namely, the decimal expressions). For instance,

$$\lfloor \frac{100}{3} \rfloor = \lfloor 33.33 \cdots \rfloor = 33$$

1

(a) For how many values of n between 1 and 40 is $\frac{n}{7}$ an integer?

(b) For how many values of n between 1 and 100 is $\frac{n}{8}$ an integer?

2 Between what two consecutive integers is $\frac{53}{5}$?

Compute $(1+3+5+7+9+11)/2$.

Solution
Let's think about the quickest way. Since addition is commutative, the order does not matter. This means that we can add numbers in any order we want. As you notice from the summation of these numbers, it is easy to compute the addition as $(1+11)+(3+9)+(5+7) = 12+12+12 = 36$. Hence, the answer must be $36/2 = 18$.

In fact, if we deal with subtraction, if we change $a-b$ into $a+(-b)$, then we can use the commutative property. For instance, $3-7+17-13 = (17-7)+(3-13) = 10-10 = 0$.

3 Given $\dfrac{a}{p}+\dfrac{b}{p}$, we can always turn the expression into $\dfrac{a+b}{p}$. Hence,

$$\frac{1}{7}+\frac{5}{7}+\frac{9}{7}+\frac{13}{7}+\frac{17}{7}+\frac{21}{7}+\frac{25}{7}$$

can easily turn into

$$\frac{1+5+9+13+17+21+25}{7}$$

Simplify the expression above into the most simplified form.

Also, we can evaluate the rational expressions by substituting the numerical values, which is known as <u>evaluation</u>, as we did in section 1.4. Let's have a review of what evaluation means. It means that we consider letters as placeholders and substitute given numerical values inside the letter expressions.

Example

Evaluate $1/(2x) + 3y$ if $x = 1$ and $y = 0$.

Solution
Evaluation of a mathematical expression containing a fraction should not make us fear the expression. We simply put the numbers into the placeholders, i.e., variable expressions. Hence,

$$1/(2x) + 3y = 1/(2 \cdot 1) + 3(0) = 1/2 + 0 = 1/2$$

4 When there is a fractional expression with variables, we can evaluate it, provided that the numbers we put inside it are nice. Specifically, $\dfrac{a}{b}$ automatically implies $b \neq 0$. Remember the following rules, given $\dfrac{p}{q} \in \mathbb{R}$,[1]

- It is okay to have $p = 0$.

- It is NEVER okay to have $q = 0$.

That being written, if $x = -7$ and $y = 5$, find the value of $2xy - \dfrac{x}{2y}$.

---

[1]The symbol $\in$ can be translated into is an element of and the symbol $\mathbb{R}$ refers to the set of real numbers.

4.2 Multiplying Fractions

We usually find two types of fraction multiplications. The first one is the product of integers and fractions. The other one is the product of two pure fractions.

- Definition of division: If y is not zero, then $\dfrac{x}{y} = x \cdot \dfrac{1}{y}$.

- Reciprocal of product: If x and y are not zero, then $\dfrac{1}{xy} = \dfrac{1}{x} \cdot \dfrac{1}{y}$.

Example

Compute $\dfrac{2}{3} \times \dfrac{3}{4}$.

Solution
Just as we can see from the bullet points above,

$$\frac{2}{3} \times \frac{3}{4} = \frac{2 \cdot 3}{3 \cdot 4}$$
$$= \frac{6}{12}$$
$$= \frac{1}{2}$$

5 Compute each of the following products.

(a) $\dfrac{5}{6} \cdot \dfrac{11}{7}$

(b) $\dfrac{1}{5} \cdot (-75) \cdot \dfrac{2}{3}$

(c) $\left(-\dfrac{80}{7}\right) \cdot \left(\dfrac{14}{9}\right) \cdot \left(-\dfrac{63}{16}\right)$

6 Due to the commutative property of multiplication,

$$\frac{a}{b} \times \frac{c}{d} = \frac{a \times c}{b \times d}$$
$$= \frac{a \times c}{d \times b}$$
$$= \frac{a}{d} \times \frac{c}{b}$$

Hence, find an integer that equals the following fraction

$$\frac{27 \cdot 25 \cdot 22 \cdot 20}{3 \cdot 4 \cdot 5 \cdot 11}$$

7 Compute $\dfrac{1 \cdot 3}{6 \cdot 9} \times \dfrac{6 \cdot 9 \cdot 12}{1 \cdot 3 \cdot 5}$.

If there is an expression such as $\frac{\spadesuit}{\clubsuit}$ of $\triangle$, then we simply translate of as $\times$. In other words,

$$\frac{\spadesuit}{\clubsuit} \text{ of } \triangle = \frac{\spadesuit}{\clubsuit} \times \triangle$$

8 What number is $\frac{4}{5}$ of $\frac{5}{12}$ of 123456?

9 Assume x, y, and z are non-zero integers. If x is divided by y, the result is 7/15. If y is divided by z, the result is 12/19. What is the result when x is divided by z?

4.3 Dividing by Fractions

Recall that we define division in terms of multiplication:

$$a \div b = a \cdot \frac{1}{b}$$

In this section, we combine this definition with the rule for multiplying fractions to learn how to divide by fractions. In particular, the quotient of a/b by c/d is equivalent to the product of a/b and d/c, since

$$\frac{1}{c/d} = \frac{d}{c}$$

Example

Simplify $\dfrac{14/3}{5/4}$.

Solution

$$\frac{14/3}{5/4} = \frac{14}{3} \div \frac{5}{4}$$
$$= \frac{14}{3} \times \frac{4}{5}$$
$$= \frac{56}{15}$$

10 Simplify $\left(-\dfrac{5}{6}\right) \div \left(-\dfrac{12}{7}\right)$.

Simplify $\dfrac{1}{2} \div \dfrac{4}{5}$.

Solution

$$\dfrac{1}{2} \div \dfrac{4}{5} = \dfrac{1}{2} \times \dfrac{5}{4}$$
$$= \dfrac{1 \cdot 5}{2 \cdot 4}$$
$$= \dfrac{5}{8}$$

11 As mentioned in the previous examples, dividing by a fraction EQUALS multiplying by its reciprocal, i.e.,

$$\dfrac{A}{B} \div \dfrac{C}{D} = \dfrac{A}{B} \times \dfrac{1}{C/D} = \dfrac{A}{B} \times \dfrac{D}{C}$$

Use this to solve the following problem. Multiplying a number by 3/4 and then dividing the result by 3/5 has the same effect as multiplying the original number by what number?

4.4 Raising Powers

If n is a positive integer and $\clubsuit \neq 0$, then

$$\left(\frac{\spadesuit}{\clubsuit}\right)^n = \underbrace{\left(\frac{\spadesuit}{\clubsuit}\right) \times \left(\frac{\spadesuit}{\clubsuit}\right) \times \cdots \left(\frac{\spadesuit}{\clubsuit}\right)}_{n \text{ copies of } \spadesuit/\clubsuit} = \frac{\spadesuit^n}{\clubsuit^n}$$

Example

Simplify $\left(\dfrac{3}{2}\right)^2$.

Solution

$$\left(\frac{3}{2}\right)^2 = \frac{3}{2} \times \frac{3}{2}$$
$$= \frac{3^2}{2^2}$$
$$= \frac{9}{4}$$

12 Compute the following power expressions.

(a) $\left(\dfrac{2}{3}\right)^{-2}$

(b) $\dfrac{1}{(1/3)^5}$

(c) $\dfrac{(2/9)^2}{(8/3)^4}$

Example

Simplify $\left(\dfrac{3}{2}\right)^2 \times \left(\dfrac{2}{3}\right)^2$.

Solution

$$\left(\frac{3}{2}\right)^2 \times \left(\frac{2}{3}\right)^2 = \frac{3}{2} \times \frac{3}{2} \times \frac{2}{3} \times \frac{2}{3}$$
$$= \frac{3^2}{2^2} \times \frac{2^2}{3^2}$$
$$= \frac{9 \cdot 4}{4 \cdot 9}$$
$$= \frac{36}{36}$$
$$= 1$$

13 Compute $\left(\dfrac{7}{13}\right)^2 \left(\dfrac{13}{7}\right)^4 \left(\dfrac{7}{13}\right)^2$.

4.5 Adding or Subtracting Fractions

We perform fraction addition and subtraction by writing the fractions with a common denominator, and then applying the distributive property. As we covered in the previous problem, that is, if P is nonzero, we have

$$\frac{a}{P} + \frac{b}{P} = \frac{a+b}{P} \qquad \frac{a}{P} - \frac{b}{P} = \frac{a-b}{P}$$

We sometimes write the sum of a positive integer and a fraction between 0 and 1 as a mixed number(sometimes called as a mixed fraction), which consists of the integer immediately followed by the fraction. For example, $3 + \frac{1}{3}$ can be written as $3\frac{1}{3}$.

14 Evaluate each of the following in simplest form.

(a) $\dfrac{1}{3} + \dfrac{1}{4}$

(c) $-\dfrac{2}{3} + \dfrac{7}{15}$

(b) $\dfrac{15}{24} - \dfrac{800}{1400}$

(d) $\dfrac{4}{3} - \dfrac{13}{4} + \dfrac{7}{6}$

Which integer is closest to $\dfrac{2n+1}{n}$ for large values of n?

Solution

$$\frac{2n+1}{n} = \frac{2n}{n} + \frac{1}{n}$$
$$= 2 + \frac{1}{n}$$
$$\approx 2$$

As n gets even larger, $\dfrac{2n+1}{n}$ approaches the value of 2 even more closely.

15 Notice that, given a large natural number N,

$$\frac{N \pm 1}{N} \approx \frac{N}{N} \approx 1$$

Hence, find the integer closest to $\dfrac{127}{128} + \dfrac{33}{32}$.

1 Evaluate the following expressions.

(a) $\dfrac{3}{7} + \dfrac{4}{9}$

(b) $\dfrac{4}{5} - \dfrac{3}{10}$

(c) $\dfrac{12}{7} - \dfrac{22}{7}$

(d) $\dfrac{2}{9} \times \dfrac{3}{10}$

(e) $\dfrac{1}{5} \div \dfrac{3}{25}$

(f) $\dfrac{2}{3} - \left(-\dfrac{3}{5}\right)^{-2}$

(g) $\dfrac{3}{8} \times \dfrac{2}{3}$

2 Evaluate $\dfrac{2+x(3+5x)-3^2}{5+x+x^2}$ when $x=-3$.

3 What is the value of $-\dfrac{1}{2} \times 4 \times \dfrac{1}{8} \times 16 \times \dfrac{1}{32} \times 64 \times \dfrac{1}{128} \times 256 \times \dfrac{1}{512} \times 1024$?

4 A woman begins her work at 10:20 a.m. and estimates that it will take $8\frac{7}{15}$ hours to finish the work. At what time does she expect to finish it?

5 Compute the following product of 2016 fractions.

$$\frac{4}{3} \times \frac{5}{4} \times \frac{6}{5} \times \frac{7}{6} \times \cdots \times \frac{2019}{2018}$$

6 Simplify the following fraction expression.

$$\left(\frac{1}{2} \cdot \frac{2}{3} \cdot \frac{3}{4}\right) \div \left(\frac{6}{8} \cdot \frac{6}{9} \cdot \frac{6}{12}\right)$$

7 Compute the following expression.

$$1\left(1 - \frac{1}{1}\right) + 2\left(1 - \frac{1}{2}\right) + 3\left(1 - \frac{1}{3}\right) + 4\left(1 - \frac{1}{4}\right) + \cdots + 10\left(1 - \frac{1}{10}\right)$$

8 Evaluate the following expression by using the distributive property.

$$\frac{7}{19} \times 107 - \frac{7}{19} \times 69$$

9 Express the following fraction in the simplest form.

$$\frac{9}{5}\left(3\frac{1}{3} \cdot \frac{1}{4} - \frac{10}{12} \cdot \frac{1}{8}\right)$$

10 Evaluate the following fraction expression.

$$\frac{7}{14} \times \frac{13}{44} + \frac{7}{14} \times \frac{17}{44} + \frac{7}{14} \times \frac{27}{44} + \frac{7}{14} \times \frac{31}{44}$$

1

(a) $\dfrac{55}{63}$ (b) $\dfrac{1}{2}$ (c) $-\dfrac{10}{7}$ (d) $\dfrac{1}{15}$ (e) $\dfrac{5}{3}$ (f) $-\dfrac{19}{9}$ (g) $\dfrac{1}{4}$

2 $\dfrac{29}{11}$

3 -32

4 $6:48$ p.m.

5 673

6 1

7 45

8 14

9 $\dfrac{21}{16}$

10 1

Topic 5

Equations and Inequalities

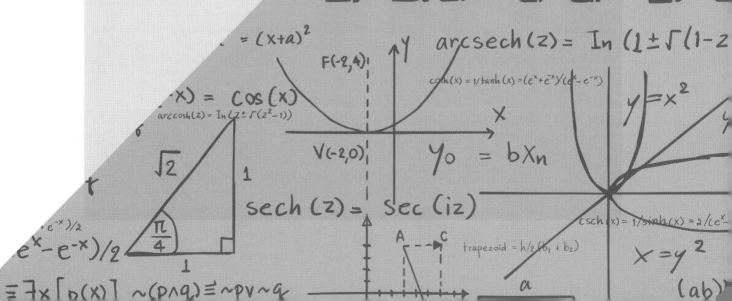

5.1 Expressions

A combination of numbers and variables using the usual operations results in a mathematical expression. For example, the following are all expressions:

$$3 + 4x - 2 \qquad 3 + 5y - 6 \qquad x^2 - 2x + 3$$

Here, a term is the product of a number and a variable raised to some power. The number in a complex-looking term is called the coefficient of the power of the variable. For example, in the expression $5x^2 + 2x$, the terms are $5x^2$ and $2x$, the coefficient of x^2 is 5, and the coefficient of x is 2.

On the other hand, a number by itself is a term as well, so in the expression $3x + 7$, both $3x$ and 7 are terms. In particular, the term that is just a number by itself is called a constant or constant term.

Two one-variable expressions with the same variable are equivalent if they are equal for any value of the variable. For example, the expressions

$$5 + 3x \qquad \text{and} \qquad 3x + 5$$

are equivalent, as are the expressions

$$\frac{x - 1}{x} \qquad \text{and} \qquad 1 - \frac{1}{x}$$

We say that we simplify an expression when we write it as an equivalent expression with as few terms as possible, and write each term as simply as possible. For example, the expression $t \cdot t + 1 + 2 \cdot 4$ can be simplified to $t^2 + 9$, but the expression $x + 7$ is already simplified.

1 Amy has five packs of gum and Bob has six packs of gum, all of which are new and unopened. If each new pack of gum has x pieces of gum, then how many pieces of gum do they have in total?

2 Albert has three unopened boxes of candies and seven extra pieces of candies. On the other hand, Robert has four new boxes of candies and five extra pieces of candies. If each unopened box contains x pieces of candies, and Albert and Robert have the same number of candies, then what is the value of x?

3 Express the following expression as a single fraction written in k.

$$\frac{17k+4}{7} - \frac{3-5k}{5}$$

Bob wants to lose his weight by weight training. Set up the correct expression of Bob's weight if Bob loses 0.5 kilogram per day, and his initial weight is 80 kilograms. Use the expression to find out his weight after 10 days of workout.

Solution
Let t be the number of days passed since the beginning of the workout. The correct expression for Bob's weight is $B(t) = 80 - 0.5t$. We wish to figure out his weight after 10 days of workout, so we substitute $t = 10$. Hence, $B(10) = 80 - 5 = 75$ kilograms.

$\boxed{4}$ In a hypothetical world of extreme diet and calorie burning, Bob currently weighs $(29 - 2x)$ kilograms. Each day, he gains $(2x + 1)$ kilograms but loses $(3 - x)$ kilograms. How much does Bob weigh in kilograms, after three days from now on, in terms of x?

5.2 Solving Basic Linear Equations

An equation states that two quantities are equal. The most basic type of equation comes from arithmetic. For example,

$$5+3=4+4$$

This is always true, so we call it identity.

In this section, we introduce equations with a variable such that the equation is true for only some values of the variable. Unfortunately, we use the same symbol, "=", to mean that two expressions are equivalent and to write equations that are only true for some values of a variable.

For example, the equation $x+3=5$ does not tell us that $x+3$ is 5 for all values of x. If $x=3$, then $x+3$ is 6, not 5, so the equation $x+3=5$ is not true when $x=3$. However, if $x=2$, then $x+3$ is 5, so the equation $x+3=5$ is true when $x=2$. The solutions to an equation are the values of the variables that make the equation true. So, $x=2$ is a solution to $x+3=5$.

The following strategies allow us to solve a given linear equation of x.

- We simplify either the leftside, the rightside, or both sides of the equation to isolate the variable expressions. For example, in the equation

$$6x-5x+3=14$$

  we can simplify the left-hand side to $x+3$, so the equation becomes

$$x+3=14.$$

- We keep trying to isolate the variable expression so that we get a variable term on one side of the equation. For example, starting with the equation $x+3=14$, we can subtract 3 from both sides of the equation to get

$$x+3-3=14-3.$$

  Simplifying both sides of the equation then gives $x=11$, and we have found the solution to the equation. Looking back to the original equation, $6x-5x+3=14$, we see that when we have $x=11$, we get $5\cdot 11-4\cdot 11+3=14$, which is indeed a true equation.

We often solve equations with one variable by performing operations on both sides of the equation, simplifying expressions until the variable is alone on one side of the equation. When we do this, we say that we isolate the variable.

In this section, we focus on solving linear equations directly or indirectly given to the readers. For instance,

$$x + 2x - 5 = 3 - 4x \qquad \text{and} \qquad 5y + 2 = 1 - 4y$$

are linear equations. The equations

$$x^2 = 49 \qquad \text{and} \qquad \frac{2}{x^3} = 19$$

are not linear equations.

Example

Determine whether $2/x - 3/y = 1$ is a linear equation of x and y.

Solution
No, they are not linear equations of x and y. An equation is a linear equation if every term in the equation is a constant term or is a constant times the first power of the variable.

Let's look at the following example to see how we isolate the variable to solve linear equations in one variable.

Example

Mr. Yoo, a math teacher, asked Bob's class to solve the equation $5x - 6 = 4x + 9$. Bob copied the coefficient of x on the right-hand side incorrectly and wrote down a different equation. Bob figured out that his equation had no solutions. What was the wrong coefficient of x on the right-hand side of Bob's equation?

Solution
If Bob uses any coefficient besides 5 on the right-hand side, he will be able to put all the variable terms on one side and solve for the variable. But if he uses 5 as the coefficient of x on the right-hand side, then he will get $5x - 6 = 5x + 9$. Subtracting $5x$ from both sides gives $-6 = 9$, which clearly has no solutions. So, Bob must have used $\boxed{5}$ as the coefficient of x on the right-hand side.

---

5 If $2019 - 2018 + 2017 = 1989 - 1990 + 1991 + N$, what is the value of N?

6 If $\dfrac{1}{n} \cdot \dfrac{1}{2} \cdot \dfrac{1}{3} \cdot \dfrac{1}{5} \cdot \dfrac{1}{7} = \dfrac{1}{2} \cdot \dfrac{1}{4} \cdot \dfrac{1}{6} \cdot \dfrac{1}{10} \cdot \dfrac{1}{14}$, what is the value of n?

7 Solve for p if

$$5p - 3(p - 4) = 6p + 6$$

8 Solve for p, q and r where

$$\frac{7}{18} = \frac{p}{72} = \frac{p+q}{108} = \frac{r-p}{126}$$

Simplify the following expression into a single fraction.

$$\frac{3x}{4} - \frac{5x}{3}$$

Solution

As one can easily check, x is the common expression, so we can group two terms by x.

$$\frac{3x}{4} - \frac{5x}{3} = \left(\frac{3}{4} - \frac{5}{3}\right)x$$
$$= -\frac{11}{12}x$$

9 Solve for x if

$$\frac{x}{2} - \frac{3}{4} = x - \frac{5}{12}$$

5.3 Word Problems

Most word problems can be solved using the following general method.

- Read the problem carefully.

- Convert the words to math.

- Solve the math.

- Convert your answer back to words.

- Check your answer (and make sure that you answered the question that was asked).

Example

If twice the value of x equals the difference of 5 and x, what is the value of x?

Solution
If $5 > x$, then the difference between 5 and x is $5 - x$. According to the question, we can set up $2x = 5 - x$. Hence, $3x = 5$, so $x = 5/3$.

On the other hand, if $x > 5$, then the difference between 5 and x is $x - 5$. According to the question, we can set up $2x = x - 5$. Hence, $x = -5$, which contradicts the original assumption. Therefore, $x = 5/3$ is the only solution.

10 Of three positive integers, the second is twice the first, and the third is twice the second. One of these integers is 23 more than another. What is the largest integer of the three?

11 Joshua is as old now as Harry was fifteen years ago. Six years from now, Harry will be twice as old as Joshua will be then. How old is Harry now?

12 Jamie has a bag of coins containing the same number of nickels, dimes and quarters. The total value of the coins in the bag is $9.20. How many coins does Jamie have in total?

13 When a positive number x is multiplied by the sum of the number and its reciprocal, the result is two. Find the value of x.

14 A parking lot in Apgujeong allows two types of automobiles - four-wheeled cars and two-wheeled bikes. If there are twenty-seven cars and motorcycles in total, and there are seventy-two wheels altogether, how many cars are there?

5.4 Inequalities

In this section, we deal with expressions that are not equal. If we know that one expression is greater than another, we can write an inequality to show this relationship, for instance,

$$3+4 > 2$$

The $>$ symbol means "greater than," so $3+4 > 2$ tells us that $3+4$ is greater than 2. We can change this expression by putting 2 on the left-side.

$$2 < 3+4$$

The $<$ symbol means "less than," so $5 < 2+7$ tells us that 5 is less than $2+7$. Similarly, $>$ symbol means "greater than," so $3+4 > 2$ means that $3+4$ is greater than 2.

The inequalities above are strict inequalities, meaning that one side is greater than the other. The $\geq$ symbol means "greater than or equal to," so

$$3+4 \geq 2$$

means $3+4$ is greater than or equal to 2. Similarly, the $\leq$ symbol means "less than or equal to."

Just as with equations, we can include variables in inequalities, such as:

$$x > 2$$

This tells us that x is greater than 2. In the real number line, the inequality tells us the position of numbers relative to other numbers.

As shown in the figure, $y < 0 < x$. Inequality tells us whether a given number is smaller or bigger than the other.

Example

If $x+3 \leq -x+1$, what is the maximum value of x?

Solution
Isolating the variables, we get $2x \leq -2$. Dividing by 2 does not change the inequality sign, so $x \leq -1$. Hence, the maximum value of x is -1.

15 How many integers n satisfy the following inequalities?

$$4n + 3 < 25$$
$$-7n + 5 < 24$$

16 Bob the baker bakes a batch of chocolate muffins and splits the batch evenly onto six different trays. He then adds seven cookies to each tray. If the tray now contains at least twenty goods, what is the least possible number of chocolate muffins in Bob's original batch?

17 A firm α charges a \$150 initial fee plus \$29 for each cup sold. Another firm β has no set up fee, but charges \$44 per cup. What is the minimum number of cups for which a customer saves money by using α? (Assume that the customer "saves" money if he or she spends less or equal amount of money.)

18 Bob brings whole roll-cakes - chocolate and cheese flavor- to his class for his birthday. The number of chocolate roll-cakes he brings is at least 2 more than 2/3 the number of cheese cakes, but no more than twice the number of cheese cakes. Find the smallest possible value for the total number of roll-cakes he brings.

1 Simplify each of the following expressions.

(a) $2(4-3r) - \dfrac{1}{3}(6+18r)$

(b) $\dfrac{25x}{21} + \dfrac{42x}{49} - \dfrac{2x}{7}$

(c) $4x + \dfrac{x-8}{2} + \dfrac{3x}{5}$

(d) $\dfrac{25r-1}{3} - \dfrac{16r-4}{12}$

[2] Solve each of the following equations.

(a) $5x - 12\dfrac{3}{5} = x + 2\dfrac{4}{15}$

(b) $\dfrac{3}{14}t = -12$

(c) $5x - 4 = 7 - 5x + 3(2 - 5x)$

[3] For what value of x does $\dfrac{x}{3} = \dfrac{1/2}{9}$?

4 Solve for x if

$$\frac{1}{2^1} + \frac{1}{2^2} + \frac{1}{2^3} + \frac{1}{2^4} + \frac{1}{2^5} = \frac{x}{2^6}.$$

5 Bob solved the equation $2x - 7 = x/7 + 9$ and found $x = 5$. He instantly knew he made a mistake. How did he know so quickly that he made a mistake?

6 Five more than twice a number equals ten less than half the number. What is the number?

7 Bob multiplied a certain number by $3\frac{1}{3}$ correctly and got 50 as an answer. However, he was supposed to have divided the number by $3\frac{1}{3}$. What is the original answer that he should have found?

8 From a "Delicious" apple tree, Albert picked $\frac{1}{7}$ of the apples and Bob picked $\frac{1}{5}$ of the apples. Charlie picked the rest of the apples. If Bob picked 6 more apples than Albert did, how many apples did Charlie pick?

1

(a) $6 - 12r$ (b) $\dfrac{37x}{21}$ (c) $\dfrac{51x}{10} - 4$ (d) $7r$

2

(a) $x = \dfrac{223}{60}$ (b) $t = -56$ (c) $x = \dfrac{17}{25}$

3 $x = \dfrac{1}{6}$

4 $x = 62$, where $\dfrac{1}{2^1} + \dfrac{1}{2^2} + \dfrac{1}{2^3} + \dfrac{1}{2^4} + \dfrac{1}{2^5} = \dfrac{31}{32}$.

5 The left-hand side of the equation is an integer, but the other side of the equation is not an integer. Hence, Bob was able to tell it right away.

6 $x = -10$.

7 $x \times 3\dfrac{1}{3} = 50$ implies $x = 15$. Hence, $15 \div \dfrac{10}{3} = \dfrac{9}{2}$.

8 Charlie picked 69 apples.

Topic 6

Decimal Expressions

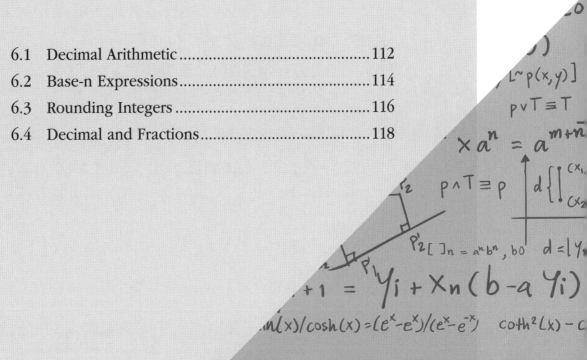

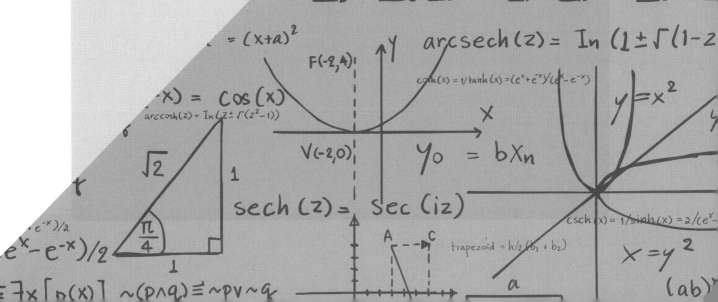

6.1 Decimal Arithmetic

Natural numbers we normally use are called the base-10 system, i.e., decimal expressions. For instance, 312 can be written as

$$3 \cdot 100 + 1 \cdot 10 + 2 = 3 \cdot 10^2 + 1 \cdot 10 + 2$$

Decimal expressions are the sum of powers of 10. We use a decimal point to distinguish the integers and decimal parts. For example, $12.23 = 12 + 0.23$ where 12 is an integer and 0.23 is a decimal expression. For instance, we can write 12.23 into

$$
\begin{aligned}
12.23 &= 10 + 2 + 0.2 + 0.03 \\
&= 1(10) + 2 + 2(10)^{-1} + 3(10)^{-2} \\
&= (1 \cdot 10^1) + (2 \cdot 10^0) + (2 \cdot 10^{-1}) + (3 \cdot 10^{-2})
\end{aligned}
$$

$\boxed{1}$ Compute the following expressions.

(a) $3.1 + 2.4$ 　　　　　(b) $12.3 + 21.45$ 　　　　　(c) $3 - 2.31$

$\boxed{2}$ Compute the following expressions.

(a) $3.56 \div 10$ 　　　　　(b) $12.34 \cdot 100$ 　　　　　(c) $0.00023 \div 100$

3 Compute the following expressions.

(a) 2.3×4

(b) 1.3×2.3

(c) 0.012×0.001

4 Compute the following expressions.

(a) $(0.12)^2$ (b) $1 \div 0.04$

6.2 Base-n Expressions

Let's look at the following table and fill up the blanks. This following example shows the 1-to-1 correspondence between base-10 and base-6 numbers.

$$63 = \underline{1} \cdot 6^2 + \underline{4} \cdot 6^1 + \underline{3} \cdot 6^0$$
$$64 = \underline{} \cdot 6^2 + \underline{} \cdot 6^1 + \underline{} \cdot 6^0$$
$$65 = \underline{} \cdot 6^2 + \underline{} \cdot 6^1 + \underline{} \cdot 6^0$$
$$66 = \underline{} \cdot 6^2 + \underline{} \cdot 6^1 + \underline{} \cdot 6^0$$
$$67 = \underline{} \cdot 6^2 + \underline{} \cdot 6^1 + \underline{} \cdot 6^0$$
$$68 = \underline{} \cdot 6^2 + \underline{} \cdot 6^1 + \underline{} \cdot 6^0$$
$$69 = \underline{} \cdot 6^2 + \underline{} \cdot 6^1 + \underline{} \cdot 6^0$$
$$70 = \underline{} \cdot 6^2 + \underline{} \cdot 6^1 + \underline{} \cdot 6^0$$
$$71 = \underline{} \cdot 6^2 + \underline{} \cdot 6^1 + \underline{} \cdot 6^0$$
$$72 = \underline{} \cdot 6^2 + \underline{} \cdot 6^1 + \underline{} \cdot 6^0$$
$$73 = \underline{} \cdot 6^2 + \underline{} \cdot 6^1 + \underline{} \cdot 6^0$$
$$74 = \underline{} \cdot 6^2 + \underline{} \cdot 6^1 + \underline{} \cdot 6^0$$

When we write numerals using digits that represent the first b whole numbers

$$0, 1, 2, \ldots, b-2, b-1$$

the numbers are written in base b, where b is the base of the number system.

5 Write each of the following sums as a single base-8 numeral.

(a) $7 \cdot 8^1 + 2 \cdot 8^0$

(b) $4 \cdot 8^2 + 3 \cdot 8^1 + 2 \cdot 8^0$

(c) $7 \cdot 8^3 + 6 \cdot 8^2 + 5 \cdot 8^1 + 4 \cdot 8^0$

(d) $5 \cdot 8^5 + 3 \cdot 8^2$

Some base number systems are common enough that we give them their own names. Here are some of the most commonly used names for base number systems:

Base	Number System
2	binary
3	ternary
4	quaternary
5	quinary
6	senary
7	septenary
8	octal
9	nonary
10	decimal
11	undenary
12	duodecimal
16	hexadecimal
20	vigesimal
60	sexagesimal

The following list of digits is commonly used for base 16:

$$0, 1, 2, 3, 4, 5, 6, 7, 8, 9, A, B, C, D, E, F$$

With these digits we can count to thirty:

$$0, 1, 2, 3, 4, 5, 6, 7, 8, 9, A, B, C, D, E, F, 10, 11,$$

$$12, 13, 14, 15, 16, 17, 18, 19, 1A, 1B, 1C, 1D, 1E.$$

Some number bases are commonly used with computers. Base 16, known also as hexadecimal, is often used to program color graphics in computers.

6 What's the largest three-digit base-9 integer? Express your answer in base 10.

6.3 Rounding Integers

Rounding is an approximation of numbers to the closest integer written in a sum of powers of 10. For example, 1320 is closer to 1300 than 1400, so we round it to 1300. What if we are in the exact middle? Then, we choose the larger one.

Example

Round 127 to the nearest hundred and ten.

Solution

First, $100 < 127 < 200$. We need to choose whether 127 is rounded down to 100 or up to 200. Since 127 is closer to 100, we round it down to 100.

Second, $120 < 127 < 130$. We will choose 130 as the round-up value of 127 to the nearest ten because 127 is closer to 130.

---

7 Round the following numbers.

(a) 867 to the nearest hundred.

(b) −2311 to the nearest thousand.

(c) 12.341 to the nearest tenth.

(d) 0.00875 to the nearest thousandth.

Example

If x rounds to 10 when rounded to the nearest tenth, find the range of the possible values of x.

Solution

If x is one-digit expression, then $5 \leq x < 9$. On the other hand, if x is two-digit expression, then $10 \leq x < 15$. Hence, $5 \leq x < 15$ is the range of possible values of x.

8 If a number x rounds to 2.9 when rounded to the nearest tenth, find the range of the possible values of x.

6.4 Decimal and Fractions

Fractions can be turned into decimal expressions in two forms :

- terminating decimals : a fraction that contains at least one power of 2 or 5.

- infinite repeating decimals : a fraction that does not contain any power of 2 and 5.

First, let's look at terminating decimals. The key idea is to make the denominator powers of 10. In other words, the only fractions that could terminate MUST contain some powers of 2 or 5 in the denominator after simplification.

Example

Convert $\dfrac{3}{4}$ into decimal expression.

Solution
$$\frac{3}{4} = \frac{3 \cdot 25}{4 \cdot 25} = \frac{75}{100} = 0.75$$

9 Express the reciprocal of 2.1 as a fraction.

10 Write the following fractions as decimal expressions.

(a) $\dfrac{3}{8}$ 　　　　　　(b) $\dfrac{2}{5}$ 　　　　　　(c) $\dfrac{19}{100}$

Many fractions (like 1/2 or 3/100) are easily converted to decimals (1/2 = 0.5 and 3/100 = 0.03). But other fractions, even though they are simple in fraction form, are quite challenging to write as decimals. Infinite decimals are divided into two parts - repeating decimals or decimals with no patterns. Decimals with no pattern are called irrational numbers. If you use long division, then we get

$$\frac{1}{3} = 0.3333\ldots = 0.\overline{3}$$

We see that the simple fraction 1/3 cannot be written as a decimal with only finitely many digits, but instead is the infinite decimal $0.333\ldots$. On the other hand, decimals that do not repeat forever, like 0.5 and 0.67676, are called terminating decimals or finite decimals.

We have a symbol that we use to write repeating decimals such as

$$\frac{2}{3} = 0.666\ldots = 0.\overline{6}$$

where the bar over the 3 indicates that the 6 repeats forever. We can have decimals in which more than one digit repeats–for example

$$0.3\overline{45} = 0.3454545\ldots,$$

where we have a single "3" followed by a "45" that repeats forever.

11 Write the following as repeating decimals.

(a) $\dfrac{5}{7}$ (b) $\dfrac{2}{9}$ (c) $\dfrac{11}{13}$

12 What is the 50$^\text{th}$ digit to the right of the decimal point in the decimal representation of $\dfrac{2}{7}$?

Every fraction can be converted into a finite decimal or infinite repeating decimal. How about infinite decimal expressions that do not repeat? These numbers are called irrational numbers. Both rational numbers and irrational numbers make up the real numbers, which can be represented as infinite dots on the number line.

> Irrational numbers that cannot be written as fractions can be categorized into two forms :
>
> - algebraic irrational numbers : $\sqrt{2}$, $\sqrt[3]{-5}$, $\cdots$
>
> - transcendental irrational numbers : π, e, $\cdots$

13 What integer is equal to $0.\overline{9}$?

1 Arrange the following numbers from smallest to largest.

$$0.59 \qquad 0.5959 \qquad 0.595959$$

2 Evaluate the product $100 \times 0.04 \times 1.44 \times 100$ without using a calculator.

3 By how much does $\dfrac{3}{2}$ exceed its reciprocal?

4 Write each decimal as a fraction.

(a) 0.35

(b) 0.23

(c) $0.\overline{37}$

(d) 2.88

(e) −1.52

(f) $3.\overline{17}$

(g) 1.875

(h) $0.25 \cdot 0.24$

(i) $(-0.5)^3$

5 Find the 150th digit to the right of the decimal point in the decimal form of

$$\frac{3}{13}$$

6 Express the following repeating decimals as a fraction.

(a) $0.\overline{13}$

(b) $0.34\overline{7}$

(c) $0.1\overline{23}$

1 $0.59 < 0.5959 < 0.595959$

2 576

3 $\dfrac{5}{6}$

4

(a) $\dfrac{7}{20}$ (b) $\dfrac{23}{100}$ (c) $\dfrac{37}{99}$

(d) $\dfrac{72}{25}$ (e) $-\dfrac{38}{25}$ (f) $\dfrac{314}{99}$

(g) $\dfrac{15}{8}$ (h) $\dfrac{3}{50}$ (i) $-\dfrac{1}{8}$

5 9

6

(a) $\dfrac{13}{99}$

(b) $\dfrac{313}{900}$

(c) $\dfrac{61}{495}$

Topic 7

Ratio and Rates

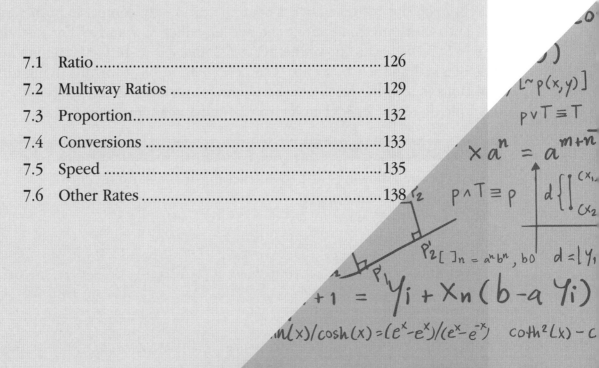

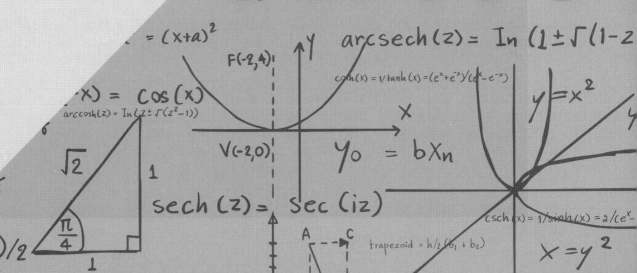

7.1 Ratio

A ratio is used to compare the quantities of data (usually in two groups). The form we use the most is

$$x : y = a : b$$

for real numbers x, y, a, and b. Ratio can also be given by fraction, i.e.

$$x : y = \frac{x}{y}$$

The key concept to remember is that the ratio only compares both x and y–it will not tell us anything about the actual values of x and y. For instance, if $x : y = 2 : 3$, we never know whether $x = 2$ and $y = 3$. In fact, we may have the case of $x = 4$ and $y = 6$. Hence, the best method of problem-solving for ratio questions is to bring forth another common variable. In the previous example, if $x : y = 2 : 3$, then let $x = 2k$ and $y = 3k$ for some k.

Example

Express the following ratio as a fraction in the simplest form.
(a) 9 pens out of 12 pens
(b) 11 feet to 25 inches

> **Solution**
> (a) Simply speaking, we may write it $9 : 12 = \frac{9}{12}$. When we factorize $\frac{9}{12}$, we get $\frac{3}{4}$.
>
> (b) Since 1 foot equals 12 inches, we have to <u>equalize</u> the units. Hence, we get $11 \times 12 = 132$ inches to 25 inches. Therefore, the ratio can be changed into the fraction
>
> $$\frac{132}{25}$$

---

1 The ratio of female to male dogs in a dog daycare facility is $4 : 5$. If there are 25 male dogs in the facility, then how many dogs are there?

2 A 10-inch length of steel is cut into two pieces whose lengths are in the ratio $2 : 3$. What is the length of the shorter piece?

3 The ratio of professors to the incoming class in a college is 1 to 11. The ratio of female freshmen to male freshmen is 4 to 5. If there are 396 female freshmen, then how many professors are there?

4 The ratio of losses to wins for Bob's basketball team is 3 to 2 for the past 10 games played. If the team additionally played the same number of games, but won half as many of the new games, then what is the changed ratio of losses to wins?

5 The ratio of black to blue marbles is 2 to 5 and there are a total of 245 marbles in the jar. How many black marbles should be added into the jar, while maintaining the number of blue marbles unchanged, to make the ratio of black to blue marbles be 3 to 7?

7.2 Multiway Ratios

A ratio is useful for comparing two quantities. How about more than two quantities? Suppose there are 5 girls and 4 boys and 1 teacher in a classroom. Assume this ratio is fixed for each classroom. In other words, we have the ratio

$$5 : 4 : 1 : 1$$

where 5 girls, 4 boys, and 1 teacher are in 1 classroom. Then, we are interested in the value of x from $x : y : z : w$. Suppose there are 6 classrooms. Then, let $x = 5k$, $y = 4k$, $z = 1k$, and $w = 1k$, for some natural numbers k. In our cases, we get $k = 6$, so we find $x = 30$, $y = 24$, $z = 6$, satisfying $30 : 24 : 6 : 6 = 5 : 4 : 1 : 1$.

Another type of multiway ratio questions asks us to simplify the given ratio. When we simplify the ratio, we always make everything into simplified integers.

6 Simplify the following ratios.

(a) $5 : 10 : 15$

(b) $12 : 16 : 4 : 8$

(c) $\dfrac{1}{2} : \dfrac{2}{3} : \dfrac{1}{3}$

7 A League of Legend, also known as LOL, tournament pays out the prizes to the top 3 teams in the ratio $5:2:1$. If the total prize money is $\$1,000,000$, then how much does the first-place team receive?

8 Bob the baker has the secret recipe for baking cakes that requires butter, flour, sugar, and milk in the ratio $1:6:2:1$, respectively. If Bob has 4 cups of sugar, how much of the other ingredients does he need?

9 There are three siblings to receive a gift card of $338 to split in the ratio of $\frac{1}{2} : \frac{1}{3} : \frac{1}{4}$. What is the greatest amount in dollars that any of the siblings will receive on his or her own?

7.3 Proportion

Whenever we have two ratios that are equal, we have a proportion. Proportion is used when we have two changing quantities with their ratio fixed. Also, probability is also the subject of proportion.

For example, suppose that Bob's secret recipe for Cafe Latte uses 5 ounces of milk and 3 ounces of coffee shots, and produces a 8-ounce glass of Cafe Latte. The ratio of milk to coffee shots is $5 : 3$. If Bob wants to make a big glass of Cafe Latte for eight friends of his, then he may require $8 \times 5 = 40$ ounces of milk and $8 \times 3 = 24$ ounces of coffee shots. No matter what quantity of Cafe Latte we want, the ratio of milk to coffee shots in Bob's recipe will always be $5 : 3$. In this case, we say that the milk and coffee shots are proportional or in proportion.

$\boxed{10}$ A not-so-accurate map of Gangnam-gu has the scale such that 1 inch $= 5$ km. If Apgujeong-dong and Daechi-dong are 4 inches apart on the map, then how far apart are the two places in terms of kilometers?

$\boxed{11}$ My wallet-size photo of my pet dog Bob is 2-centimeter wide and 4-centimeter tall. If a larger photo of Bob the lovely dog, proportional to the small photo I have, is supposed to be placed on my wall, and I want the area of the photo to be 32 square centimeters, then how many centimeters wide should the larger photo be?

7.4 Conversions

Use the following conversion ratios to solve conversion problems.

$$12 \text{ inches} : 1 \text{ foot}$$

$$3 \text{ feet} : 1 \text{ yard}$$

Example

If a car has the fuel efficiency of 14 miles per gallon, and there are 4 gallons in a gas tank to begin with, how many miles does this car can travel?

Solution

We can write down the conversion ratio in fraction to cancel out some of the units and leave the units we want. For example,

$$\frac{14 \text{ miles}}{1 \text{ gallon}} \times 4 \text{ gallons} = 56 \text{ miles}$$

12 How many yards are equivalent to 48 inches?

13 A tablespoon is half of a fluid ounce, a cup is 8 fluid ounces, and a gallon is 16 cups. How many tablespoons are in 4 gallons?

How many seconds can one hour be converted into?

Solution
Let's set up the correct fractions with units to cancel out unnecessary units.

$$1 \text{ hour } \times \frac{60 \text{ minutes}}{1 \text{ hour}} \times \frac{60 \text{ seconds}}{1 \text{ minute}} = 3,600 \text{ seconds}$$

14 The density of water is approximately 8 pounds per gallon, and there are 4 quarts in a gallon. How much does 10 quarts of water weigh?

7.5 Speed

We can write "speed is the ratio of distance to time" as the equation

$$\text{speed} = \frac{\text{distance}}{\text{time}}.$$

This equation can be rearranged as

$$(\text{speed}) \cdot (\text{time}) = \text{distance}$$

and also as

$$\text{time} = \frac{\text{distance}}{\text{speed}}.$$

Example

If Bob runs 30 minutes per mile, and Rachel runs 45 minutes per mile, who is faster?

Solution
Be careful. Look at the units carefully. This is not MILE PER MINUTES. In other words, Bob runs faster than Rachel. If we convert it to our speed, then we must set up the ratio for both runners.

Case 1. Bob's speed comes from the ratio 30 minutes : 1 mile, so he runs $\frac{1}{30}$ mile per minute.

Case 2. Rachel's speed comes from the ratio 45 minutes : 1 mile, so she runs $\frac{1}{45}$ mile per minute.

15

(a) How far does a car with a constant speed of 75 kilometers per hour travel in 2 hours?

(b) If a truck travels at a constant speed of 300 kilometers in 4 hours, then at what average speed does the truck travel?

(c) How long will a motorcycle traveling at 120 kilometers per hour need to travel 480 kilometers?

16 Bob, a casual runner, jogged for 45 minutes at a rate of 2 kilometers per hour and ran for 30 minutes at a rate of 6 kilometers per hour. How many kilometers did Bob travel at the end of 1 hour and 15 minutes? What was his average speed for the journey?

17 Bob drove 20 kilometers from his home to his office at an average speed of 80 kilometers per hour. Coming home, he encountered heavy traffic and drove the same 20 kilometers at an average speed of 40 kilometers per hour. What was his average speed for the entire 40-kilometer round trip? (Hint : The average speed is not **60** kilometers per hour.)

7.6 Other Rates

- Set up the correct rates and watch out for units.

- Manipulate the expressions to cancel out unnecessary units.

Example

Suppose Bob, an avid reader, reads a newspaper with the speed of 100 words per minute. He started reading it since $8:00$ A.M. and had read 3000 words at the moment he paused to see his watch. How many minutes must have passed at the moment he saw his watch?

> **Solution**
> "100 words $=$ 1 minutes" can be used in the following form.
>
> $$3000 \text{ words } \times \frac{1 \text{ minute}}{100 \text{ words}} = 30 \text{ minutes}$$
>
> Bob must have spent 30 minutes since 8'o clock. Hence, the time must be $8:30$ A.M.

---

18 Bob the quasi-writer can type at a rate of 50 words per minute. How long will it take him to type a 2000 word-long article, assuming that he does not stop in the middle?

19 A ultra-powerful water-hose fills a swimming pool at a rate of 1 gallon per second. If the pool's capacity is 5100 gallons, then how many hours does it take for the hose to completely fill the empty pool?

20 Bob the chatterbox wants to give a 60-minute speech. Assume that he speaks 150 elegant words per minute, and his notes contain 500 elegant words per page. How many pages should he prepare for the speech?

1 The ratio of salt to sugar of a certain mixture is $2:3$. If there are 18 grams of salt in the mixture, then how many grams of sugar are there in it?

2 The ratio of boys to girls at a Math Camp is 5 to 4. If the total number of students at the camp is 72, then how many boys and girls are at the camp?

3 Assume that one pound is sixteen ounces. Find the ratio of
1 pound, 3 ounces to 2 pounds, 5 ounces?

4 Bob the woodcutter cuts a newly-polished wooden board that is 24 meters
long into 2 pieces whose lengths have the ratio of 1 : 7. After he cuts the
wood into two pieces, what is the length of the longer piece?

5 The ratio of doctors to nurses in Gotham General hospital is 4 to 10. How many doctors are there if the number of nurses at the hospital is 150?

6 Rachel and Alex went to a shopping mall. It is well known between the two that, for every $5 Rachel spends, Alex spends $8. After they paid their lot, they noticed that Alex spent $120 more than Rachel did. How many dollars in total did they spend on the mall?

7 A marvelous stone is in average 5 kilograms, where as a dangerous stone is in average 7 kilograms. If the ratio of the number of marvelous stones to that of dangerous stones is 3 : 4 and the total weight is 86 kilograms, then how many dangerous stones are there?

8 Bob likes running. In early days of his training session, he ran the track completing 10 laps in 30 minutes. Now, he can finish 12 laps in 24 minutes on the same track. How many seconds has he improved his lap time?

9 Working alone, Jamie can paint a room in 75 minutes. If Bob helps her from the beginning of the painting work, then the two can finish painting the room in 30 minutes. How long does it take for Bob to paint the room alone?

10 Five cooks in DFAC[1] can make 20 omelets in 10 minutes. Suppose there is a to-go order for 120 omelets that should be ready in 20 minutes, then how many cooks are needed to complete the order on time, assuming that the work rate per cook does not change at all?

---

[1]DFAC is a military term for dining facility.

1 27 grams of sugar.

2 40 boys and 32 girls are at the camp.

3 19 to 37.

4 21 meters.

5 60 doctors.

6 They spend $520.

7 There are 8 dangerous stones.

8 Bob has improved his lap time by 60 seconds.

9 Bob takes 50 minutes to paint the room on his own.

10 There should be 15 cooks to complete the to-go order on time.

Topic 8

Percentages

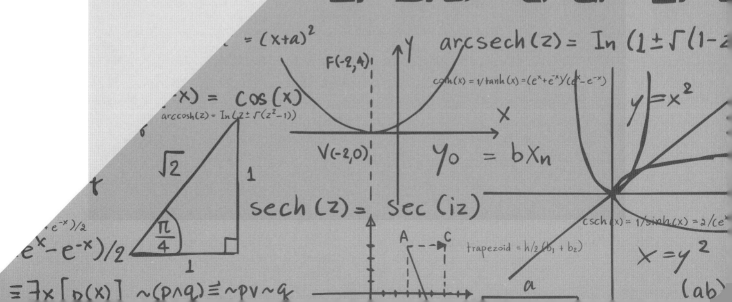

8.1 Percentages

A percent is a special type of fraction with the denominator of 100. For instance, $15\% = \dfrac{15}{100}$. Generally,

$$x\% = \frac{x}{100}$$

Example

Convert 13/20 into percentages.

Solution
Since $\dfrac{13}{20} = \dfrac{65}{100}$, the fraction represents 65%.

---

1 Write the following percents as integers, fractions, or mixed numbers.

(a) 34%   (b) 135%   (c) −420%

2 Write the following numbers as percentages.

(a) $\dfrac{23}{100}$ (b) $-4\dfrac{1}{8}$

(c) $\dfrac{1}{3}$ (d) −125

8.2 Percentage Proportion

The proportion question with respect to percentages is easily solved by making the denominator 100. Let's look at the following example.

Example

50 is what percent of 300?

Solution

Let x be the percent we want to find. Then, $\dfrac{50}{300} = \dfrac{x}{100}$. By cross-multiplication, $5000 = 300x$. Therefore, $x = \dfrac{50}{3} = 16\dfrac{2}{3}$ percentage.

3 What percent of 36 is 9?

4 15 is 20% of what number?

5 What is 35% of 40?

8.3   Word Problems

The best strategy we use is followed by

- labeling the variable(s) correctly

- setting up the correct equation(s)

- solving the equation(s)

- writing down the answer in correct context.

Example

Currently, the price of Mercedes Benz E300 is $63,000$ dollars. After two years, the car will be depreciated by 10 percent. In other words, the car price will be decreased by 10 percent. How much Bob would have to pay for the depreciated car?

Solution
Since the car must be depreciated by 10 percent, the car price must be reduced by $63,000 \times \dfrac{10}{100} = 6,300$. Subtracting it from the car price, Bob must pay $56,700$ dollars.

---

6 Bob, a tech reviewer on Youtube, bought a new smart-phone that costs $800. Where Bob lives, the sales tax is 15%. How much does Bob have to pay for this phone?

7 In the world of Harry Potter, there is a wizarding school in France called Beauxbatons Academy of Magic, enrolling 500 girls and 300 boys as its student body. What percent of the students are boys in Beauxbatons?

8 All of the students in Mr. Bob's math class took an exam. Each student either passed or failed. 80% of the students safely passed the test, but Bob was devastated to hear that five students of his class failed the exam. How many students are in the class?

8.4 Percent Increase or Decrease

Given a statement "from 10 meters to 15 meters," how would you calculate the percent of increase? The following equation will give you the tool. The equation we have is

$$\frac{\text{increased amount}}{\text{the original amount}} \times 100(\%) = \frac{5 \text{ meters}}{10 \text{ meters}} \times 100(\%) = \frac{1}{2} \times 100(\%) = 50(\%)$$

Notice that a fraction can be changed into a percentage by multiplying by 100. Similarly, the percent of decrease can be computed by similar expression, i.e.,

$$\frac{\text{decreased amount}}{\text{the original amount}} \times 100(\%)$$

Be careful that the percent of increase and decrease may not be same even if we compare the same numbers. Check out that the numbers are different in the following two questions.

9 Find the percent increase from \$20 to \$25.

10 Find the percent of decrease from \$25 to \$20.

11 State whether the following change falls under percent increase or decrease. Then, find the percent of change.

(a) From 10 to 13

(d) From 13 to 5

(b) From 40 to 33

(e) From 4 to 4.2

(c) From 20 people to 14 people

(f) From 3 to 3.5

[1] Compute 40% of 30% of 20% of 240,000.

[2] 20% of 25 is what percent of 35?

3 If there are 350 boys in a school with a total of 800 students, then what percent of the students are girls?

4 The glass gauge on a Nespresso coffee maker shows there are 80 cups left when the coffee maker is 40% full. How many cups of coffee does it hold when it is full?

5 During the baseball season, Bob the clumsy batter had 40 hits. Among his hits were 5 home runs. The rest of his hits were singles. What percent of his hits were singles?

6 Any quarter has a face value of $0.25. The collector offers to buy quarters for 200% of their face value. At that rate, how much will Bob receive for his collection of all 100 quarters?

7 Bob wants a new headphone that costs $300. Bob can buy it in his home state and pay 10% sales tax, or he can drive to a neighboring state and pay only 7% sales tax. Assuming that there is no extra cost for Bob going back and forth from his home state to the neighboring state, how much does Bob save on buying the headphone at the neighboring state?

8 200 students took "The Essential Guide to Prealgebra" course in 2019. A 30% increase in enrollment(sales) is expected each year. How many students are expected to take this online course in 2021?

1 5760

2 $\dfrac{100}{7}$%

3 56.25%

4 200 cups

5 87.5%

6 Bob will receive $50.

7 Bob will save $9.

8 338 students are expected to take The Essential Guide to Prealgebra in year 2021.

Topic 9

Square Roots

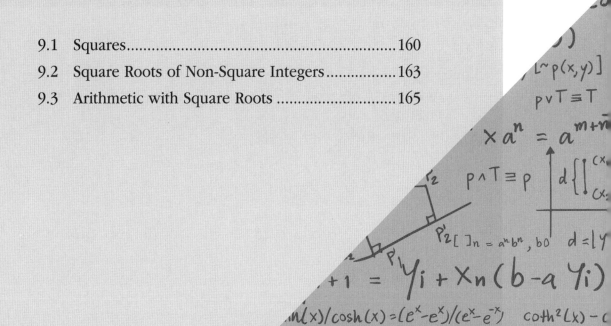

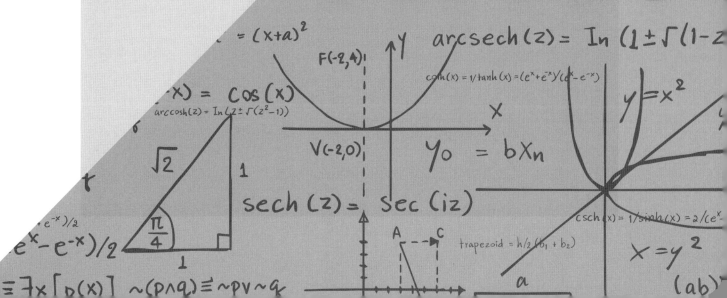

9.1 Squares

A square root of a number is one of its two equal factors. Mathematically, if $x^2 = a$ where $a \geq 0$, then x is a square root of a. For instance, 4 is a square root of 16 because $4 \times 4 = 4^2 = 16$.

The symbol we use for a square root is $\sqrt{}$, which is known as a "radical." A number[1] that goes inside the radical sign is known as "radicand." Let's learn how to read a radical number.

Example

Read the following radical numbers properly.
(a) $\sqrt{7}$ (b) $-\sqrt{11}$ (c) $-\sqrt{13}$

Solution
(a) a positive square root of 7
(b) a negative square root of 11
(c) a negative square root of 13

1 Find all values of x for which

(a) $x^2 = 16$ (b) $x^2 = -25$ (c) $x^2 = 1024$

2 Find all values of x for which

(a) $x = \sqrt{9}$ (b) $x = -\sqrt{0}$ (c) $x = \sqrt{-4}$

---

[1] A radicand is usually non-negative. If we have a negative number inside the square root, we call it imaginary number.

3 Evaluate each of the following square roots.

(a) $\sqrt{225}$

(b) $\sqrt{144}$

4 Evaluate each of the following square roots.

(a) $\sqrt{14^2}$

(c) $\sqrt{(-121)^2}$

(b) $\sqrt{123456^2}$

(d) $\sqrt{5^6}$

5 Simplify each of the following.

(a) $\sqrt{(5 \cdot 35 \cdot 7)^2}$

(b) $\sqrt{64 \cdot 49}$

(c) $\sqrt{360000}$

6 Simplify $\sqrt{8^7 + 8^7 + 8^7 + 8^7 + 8^7 + 8^7 + 8^7 + 8^7}$.

9.2 Square Roots of Non-Square Integers

What is an irrational number? Irrational number is a number that is not rational. In other words, a number that cannot be written as a quotient of integers is called irrational. What defines a rational number, then? We learned that a rational number is a fraction, or a decimal expression that is finite or infinitely progressing with repeated patterns. There are two types of irrational numbers : algebraic and transcendental. Algebraic irrational numbers are radical expressions such as $\sqrt{5}$ or $\sqrt{7}$, and we call them algebraic because they are the zeros(roots) of algebraic equations such as $x^2 - 5 = 0$ or $x^2 - 7 = 0$. Given any algebraic equation, the zero is the value of x that makes the equation 0. For example, given an equation $x - 15 = 0$, 15 is the zero of the equation.

On the other hand, transcendental number includes π or e. Transcendental number cannot be recovered by any algebraic equation with integer coefficients. Worse, the decimal expression of irrational number is infinite without patterns. In this section, we cover square roots of a non-square integer, which looks ugly in decimal expression. Nevertheless, we are expected to estimate a given square root to the nearest whole number. Oftentimes, it is best for us to compare integers with square roots.

Suppose we are interested in the integer part of $\sqrt{134}$. Then, we are looking for an integer whose square is close to 134. The closest number is 12, since $12^2 = 144$. We can easily estimate that $11 = \sqrt{121} < \sqrt{134} < \sqrt{144} = 12$.

Example

Estimate $\sqrt{231}$ to the nearest whole number.

Solution
Since $15^2 = 225$, we know that $\sqrt{231}$ is greater than 15. On the other hand, $16^2 = 256$, $\sqrt{231}$ is smaller than 16. Notice that 231 is closer to 225 than it is to 256. Therefore, we round it down to 15.

7 Compare $\sqrt{3}$ and 1.5.

Rewrite the first five two-digit natural numbers in square roots.

Solution

- $10 = \sqrt{100}$
- $11 = \sqrt{121}$
- $12 = \sqrt{144}$
- $13 = \sqrt{169}$
- $14 = \sqrt{196}$
- $15 = \sqrt{225}$

8 How many integers are between $\sqrt{15}$ and $\sqrt{142}$?

When we compare two radical expressions, always get rid of the square roots by taking a square to both side of the expressions, and use

If $a < b$, then $\sqrt{a} < \sqrt{b}$.   If $\sqrt{a} > \sqrt{b}$, then $a > b$.

9 Which is larger, $6\sqrt{11}$ or $5\sqrt{13}$?

9.3 Arithmetic with Square Roots

Square roots satisfy the following properties for all non-negative numbers a and b.

- If $a \geq 0$, then $(\sqrt{a})^2 = a$.

- If $a \geq 0$, then $\sqrt{a^2} = a$.

- If $a, b \geq 0$, then $\sqrt{a} \cdot \sqrt{b} = \sqrt{ab}$.

- If $a \geq 0$ and $b > 0$, then $\sqrt{\dfrac{a}{b}} = \dfrac{\sqrt{a}}{\sqrt{b}}$.

10 For what integer n is $\sqrt{16} \cdot \sqrt{36} = \sqrt{n}$?

11 Compute $\left(\sqrt{7} \cdot \sqrt{5} \right)^2$.

12 For what integer n is $\sqrt{5} \cdot \sqrt{7} = \sqrt{5n}$?

If a, b are non-negative real numbers, then $\sqrt{a} \times \sqrt{b} = \sqrt{ab}$. Read the following example.

Example

Evaluate $\sqrt{2} \times \sqrt{8}$.

Solution
Since 2 and 8 are positive real numbers, $\sqrt{2} \times \sqrt{8} = \sqrt{2 \cdot 8} = \sqrt{16} = 4$.

What if two numbers are both negatives? Then, we cannot split the square roots. On the other hand, if one of the two numbers is negative, then it is OKAY to split the radicals. Look at the following example.

$$\sqrt{-2} \times \sqrt{-3} \neq \sqrt{(-2)(-3)} \qquad \sqrt{-1} \times \sqrt{6} = \sqrt{-1 \times 6}$$

13 Compute each of the following expressions.

(a) $\sqrt{3} \cdot \sqrt{12}$

(b) $\sqrt{18} \cdot \sqrt{32}$

(c) $(3\sqrt{3}) \cdot (5\sqrt{27})$

14 Compute each of the following expressions.

(a) $\sqrt{\dfrac{36}{9}}$

(b) $\sqrt{\dfrac{27}{192}}$

(c) $\sqrt{11\dfrac{1}{9}}$

(d) $\dfrac{\sqrt{96}}{\sqrt{6}}$

(e) $\dfrac{\sqrt{112}}{\sqrt{28}}$

1 Evaluate the following expressions.

(a) $\sqrt{35 \cdot 5}$

(b) $\sqrt{2 \cdot 14 \cdot 21 \cdot 12}$

(c) $\sqrt{3 \cdot 5} \cdot \sqrt{3^3 \cdot 5^3}$

(d) $\sqrt{12} \cdot 3\sqrt{18}$

(e) $\sqrt{2} \cdot \sqrt{3} \cdot \sqrt{6}$

2 What is the value of the expression $\sqrt{x^2 + 11}$ when $x = 5$?

3 Simplify $\sqrt{25 + \sqrt{576}}$.

4 Which positive perfect cubes (other than 1) less than 125 have square roots that are integers?

5 If $x^2 = 25$, what is the sum of all possible values of x?

6 Find n if $\sqrt{n} = \sqrt{196} - \sqrt{121}$.

7 Find the value of x for $\sqrt{6+4x} = 6$.

8 Without using a calculator, find the greatest integer less than the following expression.

$$\sqrt{120} + \sqrt{80}$$

9 Simplify the following radical expression.

$$2\sqrt{20} - 2\sqrt{45}$$

1

(a) $5\sqrt{7}$ (b) 84   (c) 225   (d) $18\sqrt{6}$ (e) 6

2 6

3 7

4 64

5 0

6 9

7 $\dfrac{15}{2}(=7.5)$

8 Since $\sqrt{80}+\sqrt{120}<\sqrt{81}+\sqrt{121}=9+11=20$, the greatest integer less than $\sqrt{80}+\sqrt{120}$ is 19.

9 $-2\sqrt{5}$.

Topic 10
Angles

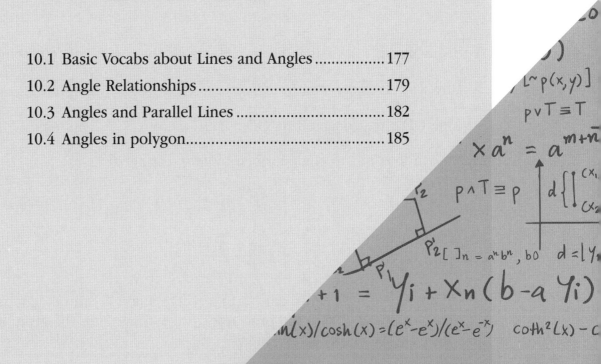

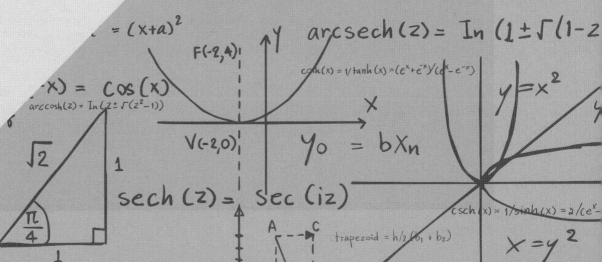

10.1 Basic Vocabs about Lines and Angles

The basic elements of geometry are point, line, and plane. First, let's learn about the basic vocabularies for lines.

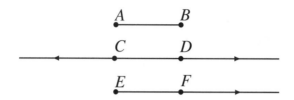

- Segment AB : It is denoted by $\overline{AB}$, where there are ____ endpoints.

- Ray EF : It is denoted by $\overrightarrow{EF}$, where there is ____ endpoint.

- Line CD : It is denoted by $\overleftrightarrow{CD}$, where there is ____ endpoint.

Now, we will look at the angle formed by two distinct rays.

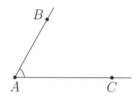

In the figure above, A is called the vertex, where angle A can be denoted by $\angle A$, $\angle BAC$, or $\angle CAB$. Using the protractor, we can measure the given angle, where there can be classified as follows.

- _____ angle : $0° < m\angle A < 90°$

- _____ angle : $90° = m\angle A$

- _____ angle : $90° < m\angle A < 180°$

- _____ angle : $180° = m\angle A$

- _____ angle : $180° < m\angle A$

1 Find the measure of x, if $m\angle AOC = 90°$, where $m\angle AOB = 3x - 10$ and $m\angle BOC = x$.

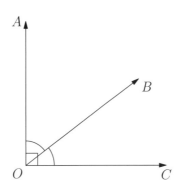

Example

For 20 minutes, how much does the hour hand move from a clock?

> **Solution**
> The hour hand moves $30°$ per one hour. Since the rotational speed of an hour hand must be constant, the hour hand must move $10°$ per 20 minutes.

2 At 2:30, what is the degree measure of the acute angle formed by the hour hand and the minute hand on a 12-hour analog clock?

3 How many degrees are in the obtuse angle formed by the hands of a clock at 3:40?

10.2 Angle Relationships

Angle relationships can be categorized into the following setups.

1. _____ Angle Pairs : two angles are _____ if the sum of the angles forms a right angle.

2. _____ Angle Pairs : two angles are _____ if the sum of the angles forms a straight angle.

3. _____ Angle Pairs : two angles are _____ if they share a vertex and a side, but no interior point.

4. _____ Angle Pairs : two angles are _____ if they are not adjacent.

5. _____ Pair : a pair of adjacent angles whose noncommon sides are opposite rays.

Given two distinct lines, angles that are opposite are called _____.

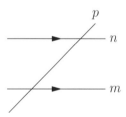

Two lines *m* and *n* are parallel, meaning that they do not cross one another inside a plane in which they are placed. Line *p*, on the other hand, cuts through two lines, usually called as a transversal. This is the usual setup for learning angle pairs, but we will see that the angle relationship still holds in nonparallel lines, as in the following problem.

Example

If two lines in the plane are parallel, and there is a transversal that passes through the two lines, then are corresponding angles congruent (= equal in measure)?

Solution
Yes. In fact, this is one of the postulates, which we take for granted, in plane geometry. We will look at more of these in the next section.

4 Identify angle relationships for the following figure.

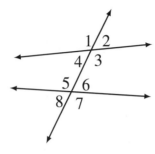

5 What is the measure of an angle, in degrees, if its supplement is five times its complement?

6 In the diagram, **BP** and **BQ** trisect ∠ABC. **BM** bisects ∠PBQ. Find the ratio of the measure of ∠MBQ to the measure of ∠ABQ.

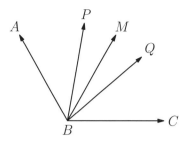

10.3 Angles and Parallel Lines

Given two parallel lines and a transversal,

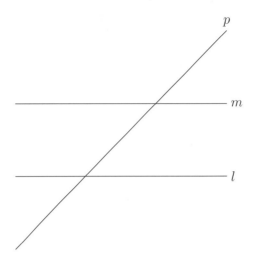

1. _____ angles are congruent.

2. _____ angles are congruent.

3. _____ angles are congruent.

4. _____ angles are congruent.

7 Prove that the sum of the angles of a triangle is always 180°.

If we continue a side of a triangle past a vertex as in the diagram, we form an exterior angle of the triangle. The interior angles of the triangle which are not adjacent to the exterior angle are called remote interior angles.

8 Prove that any exterior angle of a triangle is the sum of the remote interior angles.

9 In the diagram, A, B, and C are collinear. What is the measure of $\angle BDC$, in degrees?

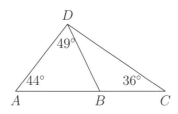

10 In the diagram, $\overline{PT}$ is parallel to $\overline{QR}$. What is the measure of $\angle PQR$ in degrees?

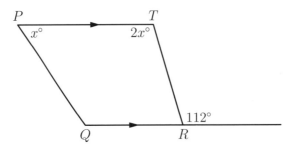

11 The trisectors of angles B and C of scalene triangle ABC meet at points P and Q as shown. Angle A measures $42°$ and angle QBP measures $15°$. What is the measure of angle BQC?

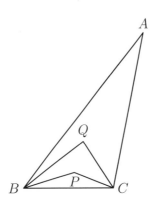

10.4 Angles in polygon

A polygon is a closed planar figure that consists of line segments. Polygons are classified by the number of sides they have, i.e., triangles or quadrilaterals. Normally, a polygon with *n* sides is generically called an *n*-gon, but many types of polygons have special names as well. The most common ones are shown in the table below.

# sides	Name
3	triangle
4	quadrilateral
5	____tagon
6	hexagon
7	____tagon
8	____tagon
9	nonagon
10	decagon
12	dodecagon

A polygon is _____ if all of its sides and all of its angles are congruent to respective sides and angles. As with quadrilaterals, any segment drawn from one vertex to a non-adjacent vertex is called a diagonal.

Figure of Regular Hexagon

$$\text{Interior angle} = \frac{180(n-2)}{n} = 180 - \frac{360}{n}$$

$$\text{Exterior angle} = \frac{360}{n}$$

12 Each interior angle of a polygon measures 170 degrees. How many sides does the polygon have?

13 A square and a regular heptagon are coplanar and share a common side $\overline{AD}$, as shown. What is the degree measure of angle *BAD*?

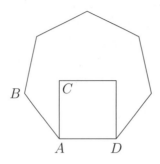

1 △*ABC* is an isosceles triangle such that *AC* = *BC*, and △*CBD* is an isosceles triangle such that *CB* = *DB*. Assume that the angle formed by $\overline{BD}$ and $\overline{AC}$ is a right angle. If *m*∠*A* = 75°, what is *m*∠*ACD*?

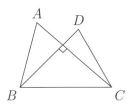

2 In △*ABC* shown, solve for *x*. (In the figure, *x* is the measure of the obtuse angle *BDA*.)

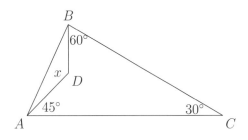

3 In the adjoining figure, *ABCD* is a square, *ABE* is an equilateral triangle and point *E* is outside square *ABCD*. What is the measure of ∠*BED*?

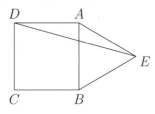

4 In △*ACD* in the figure, *m*∠*A* = 60° and *m*∠*CFD* = 110°. Assume that $\overline{CE}$ bisects ∠*ACD*(=cuts the measure of the angle into half), and $\overline{DB}$ is the altitude to $\overline{AC}$. Find the value of *m*∠*ECD*.

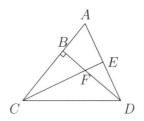

5 If the measures of interior angles of a triangle are in the ratio of $2 : 3 : 4$, what is the measure of the largest interior angle?

6 One angle of a triangle has measure $20°$ greater than another angle of the triangle and half the measure of the third angle of the triangle. Find the measure of the smallest angle.

7 What is the number of degrees in the smaller angle between the hour hand and the minute hand on a clock at 8:45?

1 $m\angle ACD = 30°.$

2 $x = 135°.$

3 $m\angle BED = 45°.$

4 $m\angle ECD = 20°.$

5 The largest angle has its measure $80°.$

6 $30°.$

7 $\dfrac{30°}{4}(= 7.56°).$

Topic 11

Perimeter and Area

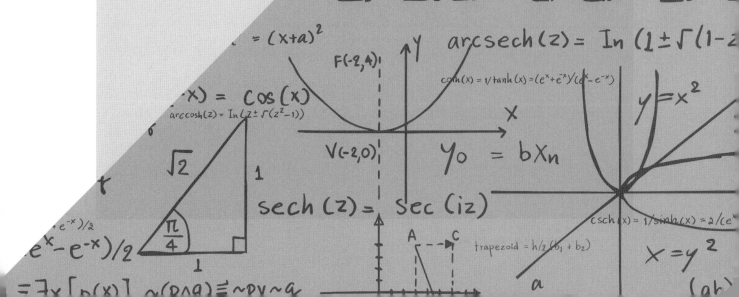

11.1 Perimeter

A usual way of measuring a closed figure is to measure the total length of the boundary, which is known as the perimeter of the figure.

In plane geometry, when we have a closed figure, it is important for us to cut it into parts that we know. Triangles are like cells of geometry. Quadrilaterals consist of two triangles, pentagons three triangles, and so on.

Let's investigate a bit more about triangles. Given a triangle whose side lengths a, b, and c, its perimeter is $a+b+c$.

- Scalene triangle : $a \neq b \neq c$

- Isosceles triangle : $a = b \neq c$

- Equilateral triangle : $a = b = c$

$\boxed{1}$ If the two triangles are equilateral triangles, where D is the midpoint of $\overline{AC}$, find the perimeter of a pentagon $\boldsymbol{ABCDE}$.

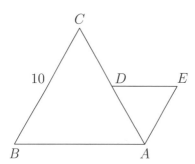

$\boxed{2}$ If an isosceles triangle has perimeter 25 where the length of each leg is twice the length of the base, what is the length of the base?

Similarly, given a rectangle whose width is w and length is l, the perimeter is $2(l+w)$.

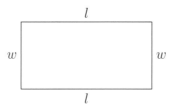

3 If a 4-foot by 8-foot rectangle is cut into four congruent rectangles with no overlapping regions, find the sum of the greatest possible perimeter of a small rectangle and the least possible perimeter of the small rectangle.

4 Rectangle *ABCD* is divided into eight squares as shown below.

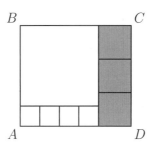

Assuming that the side length of the square that is shaded is 10 units, what is the side length of the largest square that is not shaded?

5 If each grid has a length of 1 inch, find the perimeter of the closed figures shown below.

(a)

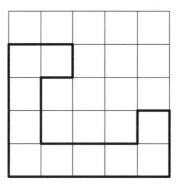

(b)

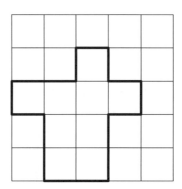

(c)

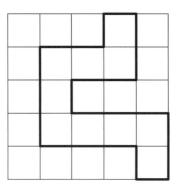

11.2 Area

While perimeter gives us a way of measuring the boundary of a closed figure, we use area to measure the space contained inside the figure.

We can think of the area of a figure as the number of unit squares that are needed to cover the figure. As you probably know, the area of a rectangle is the product of its length and width.

- Area of rectangle $= l \times w$ where l is the length and w is the width.

- Area of square $= s^2$ where s is the side length.

- Area of trapezoid $= \dfrac{1}{2}(a+b)h$ where h is height and a, b are base lengths.

- Area of parallelogram $=$ base $\times$ height , just like a rectangle.

Look at the following figure, and count the number of unit squares sitting inside the rectangle.

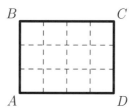

A rectangle *ABCD* has the length of 3 and the width of 4. So, the area must be $3 \times 4 = 12$ square units. Although the formula works for non-integers, the figure above explains why the formula works for integers. How many square grids can we find in the figure? There are exactly 12 square grids, each of which has the area of 1 square unit.

Example

Find out the area of a trapezoid if the measures of two bases are 3 and 4 and the height is 2.

Solution

The area must be equal to $\dfrac{1}{2}(3+4)(2) = 7$ square units.

6 Assume that the area of rectangle *ABCD* is 64 square units. Let *P* be the midpoint of $\overline{BC}$ and *Q* be the midpoint of $\overline{CD}$. What is the area of $\triangle APQ$?

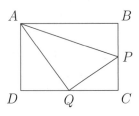

7 Assume that □*ABCD* is a rectangle whose area is 12 square units. How many square units are equal to the area of parallelogram *ECFA*?

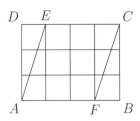

Suppose there are two triangles *ABC* and *XYZ* where $\overline{AB}$ and $\overline{XY}$ are bases. Assume that the height is equal and $AB : XY = 2 : 3$. What is the ratio of the area of *ABC* and *XYZ*?

Solution

If the height is equal, then the base ratio becomes the area ratio. Since $AB : XY = 2 : 3$, let $AB = 2x$ and $XY = 3x$ for some positive real *x*. Then, the area of *ABC* is $\frac{1}{2} \times 2x \times h$, whereas that of *XYZ* is $\frac{1}{2} \times 3x \times h$ for height *h*. Then, the area ratio is equal to

$$\frac{1}{2} \times 2x \times h : \frac{1}{2} \times 3x \times h = 2 : 3$$

which is equal to the base ratio.

8 Assume that $\triangle ABC$ has an area of 40 square units. Point *Y* is the midpoint of the segment $\overline{BX}$, where $\overline{BX}$ is the median[1] from *B* to $\overline{AC}$. What is the area of $\triangle AYX$?

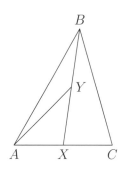

---

[1]Median in a triangle is a segment that connects a vertex to the midpoint of the opposite side.

9 If each grid has a length of 1 inch, find the area of the closed figures shown below.

(a)

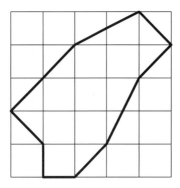

(b)

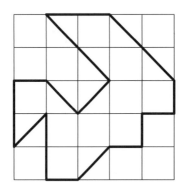

(c)

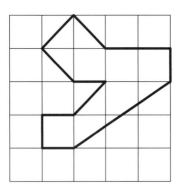

11.3 Circle

A circle consists of all points with the fixed distance, called the radius of the circle, from a given point, called the center of the circle. We also use the word "radius" to describe a segment from the center of the circle to a point on the circle. A diameter is a line segment that connects two points on the circle which passes through the circle's center. We also use the word "diameter" to mean the length of a diameter. The perimeter of a circle is called the circle's circumference.

If a circle has diameter d and radius r, then

- $d = 2r$

- The circumference is $2\pi r$.

- The area is πr^2.

Example

Find the area and circumference of a circle whose diameter is 6.

Solution
First, the radius is half the diameter, i.e., 3. The circumference of a circle is $2\pi(3) = 6\pi$ units, whereas the area of the circle is $\pi(3)^2 = 9\pi$ square units.

10 What is the radius of a circle that has the circumference of 28π units?

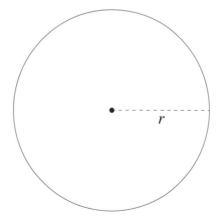

11 If the side length of a square is 10 units, find the area that is bounded by the two curves, which looks like ().

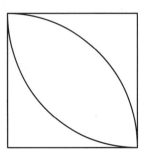

12 I have a circular garden, which is forty meters in diameter. I need one hundred trees to plant the whole garden. How many trees would I need if the diameter of the garden were tripled, still maintaining the density of planted trees in the garden?

13 Suppose one circle and three different lines are drawn on a sheet of paper. What is the largest possible number of points of intersection of these figures?

There are three figures that are similar to one another, respectively, without any reference.

Circles   Squares   Equilateral Triangles

14 If the ratio of the circumferences of two circles is 2/3, what is the ratio of their areas?

[1] Segment $\overline{AB}$ has midpoint M, and point N is the midpoint of segment $\overline{AM}$. What is the ratio of NM to AB? (How about NM to MB?)

[2] The perimeter of a quadrilateral is 100 units. One side of the quadrilateral is 28 units long and the lengths of the other sides are in the ratio $2:3:4$. Find the positive difference between the lengths of the longest and shortest sides.

3 Square tiles with 9 inches on a side exactly cover the floor of a rectangular room. The border tiles are burgundy while all the other tiles are white. The room measures 9 feet by 6 feet. How many tiles are burgundy? (Hint : 1 foot = 12 inches)

4 I start with a piece of paper that is an equilateral triangle with side length 12 inches. From each corner of this triangle, I cut away an equilateral triangle with side length 4 inches. What is the area of the figure that remains after I have removed these three pieces? (Assume the largest equilateral triangle has the area of $36\sqrt{3}$.)

5 How many different integers can possibly be the third side length of a triangle in which the other two sides have lengths 8 and 20?

6 The perimeters of two squares are in the ratio $3:5$. What is the ratio of the area of the smaller square to the area of the larger square?

7 The area of a square is 36 square centimeters. A rectangle has the same perimeter as the square. The length of the rectangle is twice its width. What is the width of the rectangle?

1. $NM : AB = 1 : 4$ and $NM : MB = 1 : 2$.

2. The longest side is 32 and the shortest is 16. Therefore, the difference is $16(= 32 - 16)$.

3. 36 tiles are burgundy.

4. $24\sqrt{3}$ is the area of the remaining figure.

5. There are 15 integers that can possibly be the side length of the triangle.

6. Two squares are similar with the length ratio of $3 : 5$. Then, the ratio of the area of smaller square to that of the larger square is $9 : 25$.

7. The width of the rectangle is 4 centimeter long.

Topic 12

Right Triangles and Quadrilaterals

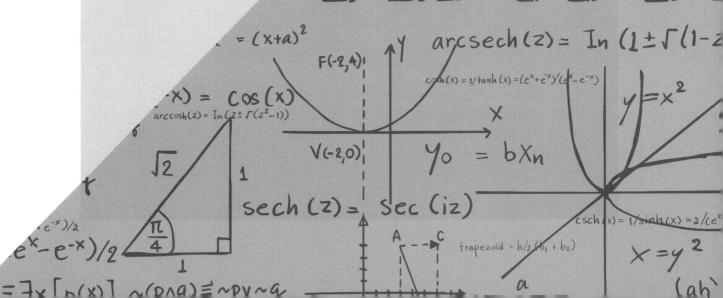

12.1 Right Triangles

Given a right triangle, the side of the triangle opposite the right angle is called the hypotenuse whereas the other sides are called the legs of the triangle.

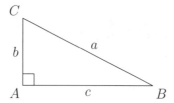

The Pythagorean Theorem states that in any right triangle, the sum of the squares of the legs is equal to the square of the hypotenuse. In our diagram,

$$a^2 = b^2 + c^2$$

There are infinitely many Pythagorean triples, such as $(3,4,5)$, $(5,12,13)$, $(8,15,17)$, and so on. Also, there are two famous special right triangles. The first is an isosceles right triangle. The legs are congruent and the hypotenuse is $\sqrt{2}$ of the leg. The following diagram shows the length ratio.

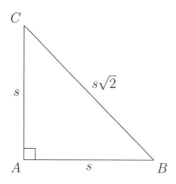

The second is called the right triangle $30 - 60 - 90$ triangle, whose side length ratio is given by $1 : \sqrt{3} : 2$.

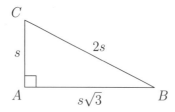

1 Assume that the perimeter of a rectangle is 28 meters, where the ratio of its length to its width is 4/3. What is the length in meters of a diagonal of the rectangle?

2 What is the length of $\overline{BC}$ of triangle ABC in the figure shown if angle D is a right angle, $AC = 13$, $AB = 15$ and $DC = 5$?

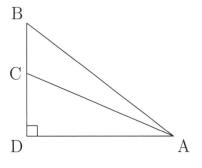

3 The hypotenuse and a leg of a particular right triangle are 4 inches and 2 inches, respectively. What is the area of the rectangle formed by attaching two right triangles of this shape?

4 Samsung or LG TV screens are rectangles that are measured by the length of their diagonals. The ratio of the horizontal length to the height in a standard television screen is $4:3$. Recently, Samsung released a new TV that has "30-inch" television screen. What is the area of the TV screen?

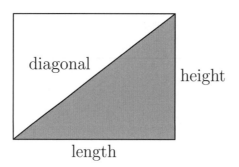

length

5 In a certain isosceles right triangle, the altitude to the hypotenuse has length $3\sqrt{2}$. What is the area of the triangle?

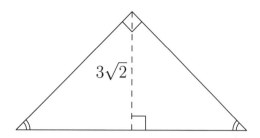

12.2 Quadrilaterals

A quadrilateral, such as the figure *ABCD* in the following question, has four segments as sides, four vertices, and four angles. Nearly all quadrilaterals we consider in this section are convex, not concave.

Convex quadrilaterals have all interior angles less than $180°$. Concave quadrilaterals, on the other hand, have some interior angles greater than $180°$. Either way, an interior angle cannot be a straight angle whose measure is $180°$. Also, the segments connecting opposite vertices are called the diagonals of a quadrilateral.

Example

Does concave quadrilateral have the sum of interior angles of $360°$?

Solution
Yes. In fact, any quadrilateral has the sum of interior angles of $360°$ because one diagonal always cuts the quadrilateral into two triangles.

6 Show that the sum of interior angles for a convex quadrilateral is $360°$, whereas the sum of exterior angles for a quadrilateral is also $360°$.

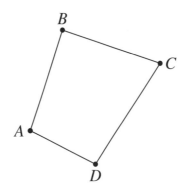

7 Bob tried to solve the following question : "The angles of quadrilateral *ABCD* satisfy $m\angle A = 2m\angle B = 3m\angle C = 6m\angle D$. What is the degree measure of $\angle A$, $\angle B$, $\angle C$, and $\angle D$?" As he wrote down the answer, he found it weird. Why did he feel odd about the solution?

12.3 Trapezoid

$$\text{Square is both} \begin{cases} \text{Rectangle} \subset \text{Parallelogram} \subset \text{Trapezoid} \\ \text{Rhombus} \subset \text{Parallelogram} \subset \text{Trapezoid} \end{cases}$$

First, we will look at a trapezoid. A trapezoid is a quadrilateral with two parallel sides. The segment connecting the midpoints of the non-parallel sides is the median(or midsegment) of the trapezoid, and the distance between the two parallel sides is the height of the trapezoid.

In an isosceles trapezoid:

1. The base angles are congruent.

2. The legs are congruent.

3. The diagonals are congruent.

Isosceles trapezoid uses symmetric properties because the legs have the same length. Let's look at the following example to see what kinds of questions can be made out of isosceles trapezoids.

Example

Given an isosceles trapezoid whose leg length is 5, the smaller base is 7, and the larger base is 13, find out the height.

Solution

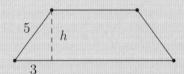

First, subtract 7 from 13. This length of 6 must be divisible by 2 because it is an isosceles triangle. Draw an altitude from the top base to the bottom base as a dashed line.

Call its length as h. Then, we form a right triangle, so it forms $3-h-5$ right triangle. Pythagorean theorem states that $3^2 + h^2 = 5^2$. Therefore, $h = 4$.

8 Quadrilateral *ABCD* is a trapezoid with $\overline{AB}$ parallel to $\overline{CD}$. We know $AB = 20$ and $CD = 12$. What is the ratio of the area of triangle *ACB* to the area of the trapezoid *ABCD*?

9 In the diagram, *PQRS* is a trapezoid with the area of 15 square units where *RS* is twice the length of $\overline{PQ}$. What is the area of $\triangle SRQ$?

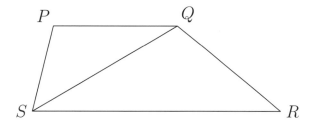

12.4 Parallelogram

The first triangle to investigate its properties is an isosceles triangle. Likewise, the first quadrilateral to investigate its properties is a parallelogram.

A parallelogram is a quadrilateral in which both pairs of opposite sides are parallel. Parallelogram satisfies the following three properties.

1. Opposite sides are congruent.

2. Opposite angles are congruent.

3. Two diagonals bisect each other.

4. One pair of opposite sides is congruent and parallel.

$\boxed{10}$ One angle of a parallelogram is 120 degrees, and two consecutive sides have lengths of 8 inches and 15 inches. What is the area of the parallelogram?

$\boxed{11}$ In parallelogram $ABCD$, $AB = 38$ cm, $BC = 2y$ cm, $CD = 2x + 4$ cm, and $AD = 24$ cm. What is the sum of x and y?

12.5 Rhombus

Rhombus is a quadrilateral with all four congruent sides. The important property we must remember is that the diagonals are perpendicular bisector of one another. When does a parallelogram turn into a rhombus?

1. Two adjacent sides are congruent.

2. Two diagonals are perpendicular.

Example

If two diagonals of a rhombus have the length of 10 and 24, find the perimeter.

Solution
Without graphing the rhombus, we know that one side must be the hypotenuse of the right triangle whose sides are half of the diagonals, i.e., 5 and 12. Hence, the side must be 13 from $5:12:13$ triangle. Therefore, the perimeter must be $4(13) = 52$ units.

$\boxed{12}$ $ABCD$ is a rhombus with diagonals $AC = 8$ and $BD = 6$. Find the area and the perimeter of $ABCD$.

If one side length of a rhombus is 8 and one diagonal length of it is 8, then find out the length of the other diagonal.

Solution

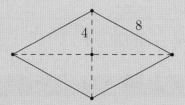

As one can check from here, the half of the unknown diagonal must be $4\sqrt{3}$ from $1:2:\sqrt{3}$ triangle. Hence, the length of the diagonal must be $2(4\sqrt{3}) = 8\sqrt{3}$.

13 If the diagonals of a rhombus measure 16 meters and 12 meters, what is the perimeter of the rhombus?

12.6 Rectangle

Rectangle is a quadrilateral with all four congruent interior angles. The important property for rectangle is that the diagonals are congruent, bisecting each other. When does a parallelogram turn into a rectangle?

1. One interior angle is right angle.

2. Two diagonals are congruent.

Example

If the width is 12 and the length is 5 for a rectangle $ABCD$, find out the length of $\overline{BD}$.

Solution
Without graphing the rectangle, we know that $\overline{BD}$ refers to the diagonal length. Since rectangle has the right angle as its interior angle, use Pythagorean theorem to find out that $BD^2 = 5^2 + 12^2 = 13^2$. Hence, $BD = 13$.

---

14 Prove that the quadrilateral formed by connecting the midpoints of a rhombus is a rectangle.

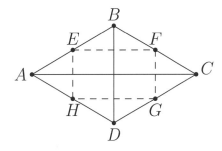

15 Five identical rectangles are arranged to form a larger rectangle *PQRS*, as shown. The area of *PQRS* is 1,500. What is the length, *x*?

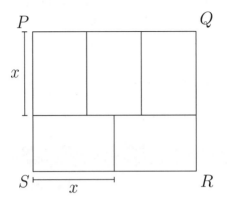

12.7 Square

Square is a parallelogram with four congruent sides and four right angles. Squares satisfy everything that is true about rectangles, rhombi, and parallelograms. If we let s be the side length of a square, P be the perimeter, and A the area. We have

$$P = 4s \qquad\qquad A = s^2$$

Drawing a diagonal creates two 45-45-90 triangles. Letting the length of a diagonal be d, we have

$$d = s\sqrt{2} \qquad P = d(2\sqrt{2}) \qquad A = \frac{d^2}{2}.$$

Example

If a square A and B has the side ratio of $5 : 7$, what is the area ratio of A and B?

Solution
Recall that squares are all similar. It means that the area ratio must be the ratio of length squared. Since the length ratio is $5 : 7$, the area ratio must be $5^2 : 7^2 = 25 : 49$.

Or, one could think of it as the specific side length, i.e., $5x$ and $7x$ for some positive real x. Then, $(5x)^2 : (7x)^2 = 25x^2 : 49x^2 = 25 : 49$.

16 If $\square ABCD$ and $\square BDFG$ are squares, then find the area ratio of

$$[BDFG]/[ABCD]$$

where $[ABCD]$ refers to the area of $\square ABCD$.

17 Given a square □*ABDF* with sides 6 centimeters shown below, if *P* is a point such that the segment $\overline{PA}$, $\overline{PB}$, $\overline{PC}$ are equal in their lengths, and the segment $\overline{PC}$ is perpendicular to segment $\overline{FD}$, what is the length of $\overline{PC}$?

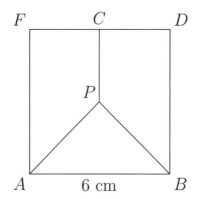

1 A rhombus is inscribed in a circle. The length of one diagonal of the rhombus is 20. What is the perimeter of a rhombus?

2 Find the area of a rhombus with a side of length 17 and one diagonal of length 30.

3 The rectangular painting is surrounded by a frame which is 4 units wide and long. If the perimeter of the outer-frame is 96 units, what is the perimeter of the painting?

4 What is the length of the common external tangent segment[1] of two externally tangent circles whose radii are 8 units and 10 units, respectively?

---

[1]External tangent line is the line that is tangent to both circles and not passing through the line segment connecting the centers of two circles.

5 The line joining the midpoints of the diagonals of a trapezoid has the length of 3 units. If the longer base is 50 units, what is the length of the shorter base?

6 Let *ABCD* be a trapezoid with the measure of *DC* three times that of base *AB*, and let *E* be the point of intersection of the diagonals. If the measure of diagonal *AC* is 12, then find the product $AE \cdot EC$.

1 $40\sqrt{2}$

2 240

3 64

4 $8\sqrt{5}$

5 44

6 27

Topic 13

Three-dimensional Figures

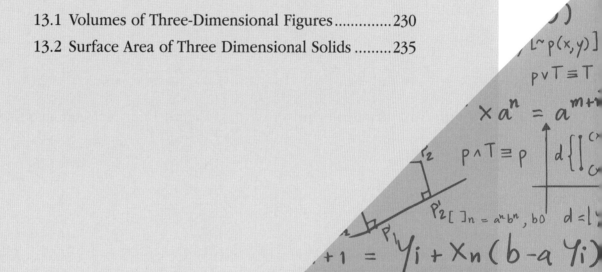

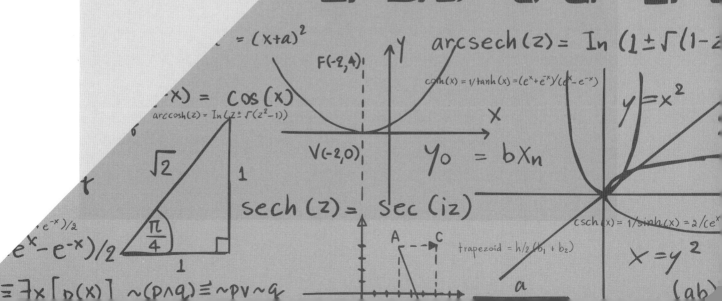

13.1 Volumes of Three-Dimensional Figures

There are three basic elements of Geometry : point, line, and plane. Here, a plane is a flat two-dimensional surface that infinitely extends. Intersection of two planes is a line. Intersection of three planes can be a point and a line. The study of multiple layers of planes leads us to the topic of three-dimensional figures.

The first type we will study is a polyhedron. Polyhedra are 3D solids whose faces are polygons. If faces are not polygons, then they are not polyhedra. There are two types of polyhedron : prism and pyramid.

Figure 13.1: triangular prism

Figure 13.2: rectangular prism

Figure 13.3: triangular pyramid

Figure 13.4: rectangular pyramid

Given a three-dimensional figure such as prism or pyramid, there are three components of the figure such as faces, edges, and vertices.

1 Identify the faces, edges, and vertices of the following triangular prism.

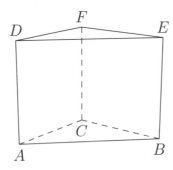

Let's compute the volume of a prism, which is the area of the base times the height, i.e.,

$$V = Bh \text{ where } B \text{ is the base area and } h \text{ is height}$$

Example

Compute the volume of the following figure.

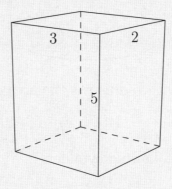

Solution
Since the base area is 3×2 and the height is 5, we compute the volume as **30** cubic units.

2 Find the volume of the following triangular prism.

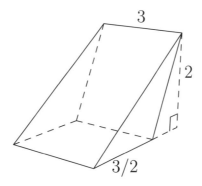

Let's look at cylinder. A cylinder is a solid whose bases are congruent circles[1], connected with a curved lateral side.

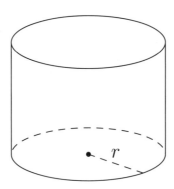

The volume of a cylinder with radius r is the area of the base times the height, i.e.,

$$V = \pi r^2 h$$

3 Find the volume of the following cylinder.

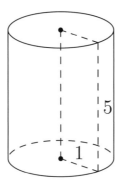

---

[1]For this reason, this is not a polyhedron.

Other than polyhedron and cylinder, there are pyramids and cones. Volume of pyramid is one third the volume of associated polyhedron.

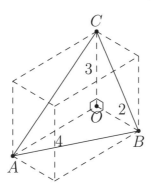

As one can see from the figure above, the volume of the pyramid *ABOC* is one third of the base area, i.e., $\frac{1}{2} \times 2 \times 4 = 4$ times the height of the pyramid, i.e., 3.

4 Find the volume of the following pyramid if *AD* = 2, *BC* = 4, *CD* = 3, and *EF* = 3.

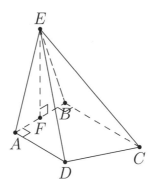

A cone (or circular cone) is a three-dimensional solid. It consists of a circular base, a point (called the vertex), and all the points that lie on line segments connecting the vertex to the base. Thus, the cone is the special case of the pyramid in which the base is circular. Similar to pyramids, the volume of cone is one third the volume of cylinder.

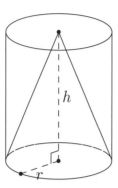

Here, the volume is given by $\frac{1}{3} \times \pi r^2 \times h$.

5 Find the volume of the following cones.

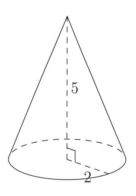

13.2 Surface Area of Three Dimensional Solids

Net is a 2-space representation of any solid. Consider a paper plane. A single sheet of paper can make a 3-dimensional paper plane that actually flies! The folding process makes a 3-space figure, whereas the folded paper is equivalent to a sheet of paper, which is a net.

The surface area of three dimensional solids is the sum of area of the faces, all of which are easily found by observing their nets. First, let's look at the surface area of prisms. Normally, the formula to find out the surface area of the rectangular prism is given by

$$S = 2wl + 2wh + 2lh$$

where w is the width, l is the length, and h is the height of the solid. Nonetheless, this formula is pointless if the base is changed into a different polygon. In fact, it is more important to count the area of all faces, rather than memorizing a formula to find the surface area.

6 Find the surface area of the following triangular prism.

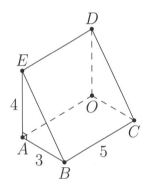

How about the surface area of a cylinder? The surface area of cylinder is easily computed by adding the area of two bases and the area of lateral face.

7 Find the surface area of the following cylinder.

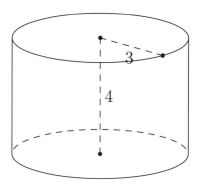

The surface area of pyramid can be computed as the sum of the lateral faces and the base face. Given the following figure, we can distinguish which is the height of the lateral face and the height of the pyramid.

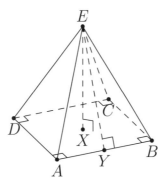

- The height of the lateral face is *EY*.

- The height of the pyramid is *EX*.

8 Find the surface area of the pyramid.

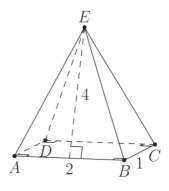

The surface area of cone can also be computed as the sum of lateral face area and the base face area.

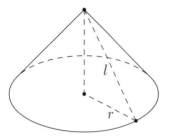

Figure 13.5: Cone

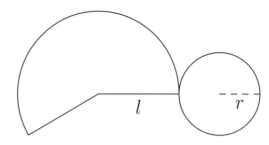

Figure 13.6: Net

9 Find the surface area of the following cone.

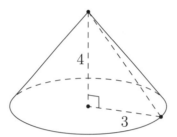

1 Find the volumes of the following prisms or cylinders.

(a)

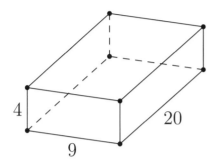

(b)

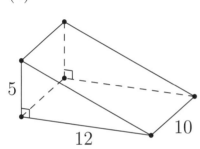

(c)

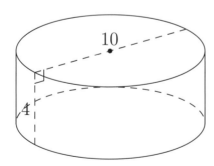

2 If $AC = 4$, $BC = 3$, and the height of the pyramid is given by **6**, find the volume of the following pyramid.

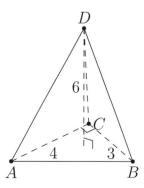

3 Find the volume of the cone if the slant height is **13** and the height of the cone is **12**.

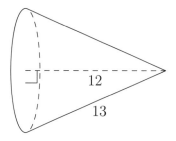

Find the surface area of the following prisms or cylinders.

(a)

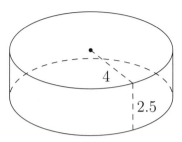

(b)

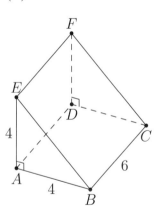

5 Find the surface area of the following pyramid or cones.

(a)

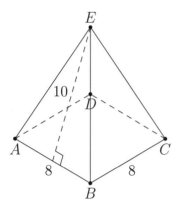

(b)

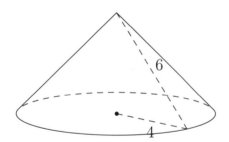

1

(a) 720

(b) 300

(c) 100π

2 12

3 100π

4

(a) 52π

(b) $64+24\sqrt{2}$

5

(a) 224

(b) 40π

Topic 14

Counting and Statistics

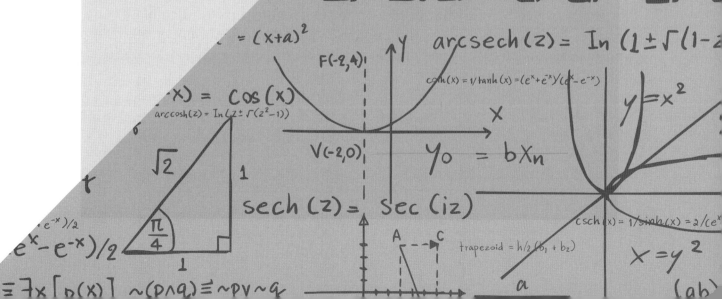

14.1 Descriptive Statistics

Suppose you took eight quizzes in Prealgebra, the scores of which are given by
$$\{90, 95, 94, 85, 87, 90, 91, 99\}$$
There are a few methods of describing the distribution of data.

- Stem-leaf plot : it is a special table where the data is split into a stem (the first digit or digits) and a leaf (the last digit).

- Box-Whisker plot : it is a graph that represents the distribution of data into five-number summary, i.e., minimum-lower quartile-median-upper quartile-maximum.

- Frequency Table : it is a list or graph that displays the frequency of data values.

- Histogram : it is a graph with frequencies to show the frequency of numerical data.

- Bar graphs : it is a graph with frequencies to compare different categories of data.

In particular, we will look at stem-leaf plots. Stem-leaf plot is used to understand the distribution of data. Descriptive statistics is the study of how data is structured. For instance, the data set $\{11, 11, 12, 20, 21, 22, 23, 33, 34, 35, 35\}$ can be represented using the stem-leaf plot by

1	112
2	0123
3	3455

Naturally, we get a box-whisker plot from the stem-leaf plots. The plot provides five numbers, i.e., minimum, lower quartile, median, upper quartile, and maximum. Median is the middle value, while the lower quartile is 25th percentile and the upper quartile is 75th percentile.

1 Display the following distribution $\{1, 2, 3, 4, 5, 6\}$ into a stem-leaf plot and box-whisker plot.

Now, let's talk about measures for descriptive statistics. There are two categories for describing the set of data.

- Measures of Variation

- Measures of Central Tendency

Meausures of central tendency are mean, median, and mode.

- Mean : the average of all data values.

- Mode : the most frequent data value.

- Median : the middle value

The following examples show the basic application of the measures of central tendency.

Example

(a) Given the set of values $\{1,2,3,4,5,6\}$, find the mean and median.

Solution
First, let's compute the mean value. The mean value is the average of all data, so

$$\frac{1+2+3+4+5+6}{6} = 3.5$$

On the other hand, the median is the $\frac{6+1}{2}$th data, which is the average of 3rd and 4th data, i.e., 3.5.

(b) Given the set of values $\{1,1,2,3,4,4,5,6\}$, find the median and mode.

Solution
First, the median value is the $\frac{8+1}{2}$th data, which is the average of 4th and 5th data, i.e., $\frac{3+4}{2} = 3.5$.

On the other hand, there are two modes, 1 and 4, so we call the data set is bimodal, meaning that there are two modes.

(c) Is there a mode in $\{1,2,3,4,5,6\}$? How about median or mean?

Solution
Every data set value appears exactly once, so we do not have a mode. Any data set should have a median and mean, but it does not necessarily have mode value(s).

Also, when we compute mean, data values in the data set don't have to be ordered from the least to greatest, but median should be computed after rewriting the data set in increasing order.

On the other hand, measures of variation follows as

- Range : the difference between maximum and minimum.

- Interquartile range : the difference between the upper quartile and the lower quartile.

Given the set of values $\{1,2,3,4,5,6\}$, find the interquartile range.

Solution

First, the median is the $\dfrac{6+1}{2}$th data, which is the average of 3rd and 4th data, i.e., 3.5. The values that are smaller than 3.5 correspond to the lower-half of the data. The median of the lower-half is called the Lower Quartile, known as LQ, which is 2, in the example.

On the other hand, the upperhalf corresponds to the set $\{4,5,6\}$, whose median is 5. This 5 is called the Upper Quartile, known as UQ. Hence, the interquartile range is the difference between UQ and LQ, $5-2=3$, in our example. Range, however, is the difference between maximum and minimum, which is $6-1=5$.

2 If five distinct positive integers have the mean and median value of 5, with a single mode of 8. Find both the range and the interquartile range.

Statistics cannot dispense with the study of probability. As a brief introduction of probability, let's look at the definition and some properties. Probability is a likelihood of an event to occur. For instance, what is the probability of raining tomorrow? Well, we can guess it as half, i.e., $\frac{1}{2}$, which is not necessarily true, but in general, it is. (It either rains or doesn't.)

The two most important properties to think about the probability are that

- $0 \le P(A) \le 1$ where $P(A) = \dfrac{\text{the number of favorable outcomes for } A}{\text{the number of total outcomes}}$.

- $P(\text{A will not occur}) = 1 - P(\text{A will occur})$.

Example

Find the probability that it will not rain today if $P(\text{it will rain today}) = 0.3$

Solution
Since $P(\text{it will not rain today}) = 1 - P(\text{it will rain today})$, we get $1 - 0.3 = 0.7$

When we learn probability for the first time, we see a new term called the sample space, the collection of all possible outcomes. For instance, when you flip a coin, you either get Head or Tail. Mathematically, we can write $U = \{H, T\}$ where U stands for the universal set, H, T representing Head or Tail, respectively.

Example

(a) Find the sample space for throwing a regular die.

Solution
Die consists of six faces, so we get $U = \{1, 2, 3, 4, 5, 6\}$.

(b) Find the probability that a multiple of 3 will show up when we throw a regular die.

Solution
Given $U = \{1, 2, 3, 4, 5, 6\}$, there are two multiples of 3, i.e., 3 and 6. Hence, the probability we want is

$$\frac{\|\{3, 6\}\|}{\|\{1, 2, 3, 4, 5, 6\}\|} = \frac{2}{6} = \frac{1}{3}$$

14.2 1-to-1 Counting

The basic counting starts with 1 and gets progressed to other integers. What if I ask you to find the number of integers between 45 and 317, exclusive? Then, we can count one by one, starting from $46, 47, \ldots, 316$. Is there a smarter way to compute this? The answer is YES. In fact, if you subtract the two and minus 1, we get the number of numbers between the given numbers. But, the fundamental process we should learn is 1-to-1 correspondence between the following two lists of numbers.

$$46, 47, \ldots, 316$$
$$1, 2, \ldots, 271$$

The first array of numbers and the second array of numbers have a 1-1 correspondence. Hence, computing the numbers from 1 to 271 is equivalent to counting numbers from 46 to 316.

3 Find the number of integers between 56 and 174 inclusive.

4 Find the number of even integers $2, 4, 6, \cdots, 432$.

14.3 Multiplication Principle

We have two major principles of counting. The first principle is $\times$, and the second is $+$. We will look at when we have to use the first principle. We multiply if

- the occurrences are repetitive. For instance, if we flip two coins, all possible results are HH, HT, TH, TT. Notice that, given any result of the first coin, either H or T, the second coin always has two results H or T. In other words, __T or __H will occur for any letter that goes inside the first underline. Hence, we multiply 2×2 where the first 2 represents all possible outcomes of the first coin, while the second 2 shows all possible outcomes of the second coin.

- the occurrences are successive. Come back to our coin flipping cases. If we imagine flipping coins separately, there must be the first and second coin to be flipped. What is the number of possible outcomes of the first coin? It is 2. Are we done? No! We must flip the last coin, too! Flip the last coin, where the number of possible outcomes of the second coin is also 2. When <u>action is ongoing</u>, we multiply the number of outcomes.

Example

Suppose there are 5 lunch menus, and 6 dinner menus. How many ways can you choose both lunch and dinner menu?

Solution
In mathematics, we use "both A and B," not as the events that occur at the same time, but the events that are successive. In our question, we eat lunch before dinner. In other words, lunch has to be done before we choose anything for dinner.

How many ways are there for us to choose a menu for lunch? There are 5 different menus, so there are 5 ways for us to choose one menu item for lunch.

Now, we will choose one menu for dinner. How many ways are there? There are 6 different menus, so there are 6 ways for us to choose one menu item for dinner.

Because all of our actions(choices) are successive(or ongoing), we simply multiply the number of choices we had, i.e., $5 \times 6 = 30$.

5 There are 26 alphabets. How many four-letter words are there?

When there are restrictions, we count the restrictions first. This is really convenient because we can multiply in any order we consider proper for the given situation.

Example

How many numbers are there with two digits?

> Solution
> Restriction is not exactly specified, but we can deduce that the number expressions with two digits do not have 0 in the tens unit.
>
> Hence, we first count the number of choices for the tens unit. We get 9 choices starting from 1 to 9 to put in the tens digit.
>
> How about the number of choices for the units digit? There are 10 choices starting from 0 to 9 to put in the units digit.
>
> We will multiply 9 and 10 because our choices are successive events. In other words, when we choose a number for the tens digit, we are not done counting. We would be done counting when we choose the units digit as well. Here, when our action is incomplete(or ongoing), we keep multiplying the number of choices.

6 How many four-letter words are there with a vowel, i.e, a, e, i, o, u, at each end, and consonants in the middle two places?

7 The original skittles come in six colors, purple, yellow, green, orange, red, and blue (limited in the UK). In how many ways can Bob distribute skittles to his five students? (Assume there are plenty amount of skittles of each color, and one student gets one skittle each.)

8 How many odd numbers with third digit 1 are there between 30000 and 69999 inclusive?

What if the number of counts depends on the priori(=previous) counts? Simple multiplication only works when all contributors are completely independent. Normally, it fails for a problem like:

In how many ways can Bob give differently-colored skittles to four students?

The answer is <u>not</u> 6^4. The reason that independence breaks down here is that there are a limited number of skittle colors. If Bob gives a red skittle to the first kid, the option no longer exists of giving a red to any of the others.

In such a situation, we can still use multiplication, but incorporating the restriction that there is only one of each color. So we line the kids up (in our head), and start giving out candy. There are 5 ways to give the first kid a piece, then only 4 for the second, since one is gone (doesn't matter which), then 3 for the third.

Example

How many ways can we select the president and the vice president out of ten students?

Solution
Let's think about the possible number of ways to choose the president out of ten students. Obviously, there are ten possibilities. Assume that the spot is filled. Then, only nine students compete for vice-presidency. Hence, the total number of counts is $10 \times 9 = 90$.

---

9 In how many ways can a 6-letter word be written using only the first half of the alphabet with no repetitions such that the third and fifth letters are vowels, and the last a consonant?

14.4 Permutation: Counting Successive Events

Permutation is the number of arrangements of k seats with n different people, denoted by $_nP_k$ and computed by

$$n(n-1)(n-2)\cdots(n-k+1), \text{ or } \frac{n!}{(n-k)!}.$$

Factorial is defined by $n! = n(n-1)(n-2)\cdots 2 \cdot 1$. There are questions that involve the concept of factorials.

10 Compute $\dfrac{5!}{4!} \times \dfrac{3!}{2!} \times \dfrac{1!}{0!}$.

11 There are 10 students in a class room. These students are running for school administration positions - class president and vice-president. How many possibilities are there for there to assign two students into the two positions?

Let's extend our argument. Permutation is an arrangement of objected in a line. If we arrange objects in a line, then we have no worries.

$$A--B--C \quad B--C--A \quad C--A--B$$

are considered distinct. But what about things laid out in a circle? How about the following figures?

A   C   B

B   C   A   B   C   A

Why are these three apparently different arrangements considered equivalent? Consider what person A sees in each case: B on the right, C on the left. To A, the arrangement looks the same in all cases. If you consider what B and C see, you will see that the three cases are equivalent to them as well.

On the other hand, reflections do matter, since after a reflection, A sees B on his or her left! Thus

A

C   B

is different from the previous three. There are only two distinct circular arrangements of 3 objects. If we count objects in a circle as we do objects in a line, we decide there are $n!$ arrangements. However, as shown above, these arrangements are not all different. Each distinct arrangement is counted n times, once for each rotation of the objects. To account for this, we must divide the number of arrangements by n, yielding $n!/n = (n-1)!$ distinct arrangements. (Compare this to our assertion that there are only two ways to arrange 3 objects.)

$\boxed{12}$ In how many ways can 3 students be seated in a round table?

14.5 Principle of Addition

The second principle of counting is +. We add if

- we have to use caseworks.

- the occurrences are not repetitive.

Example

How many ways are there to get a multiple of 3 or an odd number in a regular die.

Solution

Given $U = \{1,2,3,4,5,6\}$, let's do caseworks.

Case 1. Multiples of 3 : $\{3,6\}$

Case 2. Odd numbers : $\{1,3,5\}$

Overcounted Case : $\{3\}$ Summing up (and subtracting the over-counted number(s)), we get $2+3-1=4$.

---

13 In how many ways can Bob give one skittle each to two children if he has 3 different red, 4 different purple, and 5 different orange ones and the two children insist upon having skittles of different colors?

There are three black marbles, four green marbles, and five blue marbles. How many three marbles can be selected one at a time, without replacement, such that the first must be blue marble?

Solution
The first marble must be blue. Let's perform caseworks.

Case 1. Blue - Black - Green : $5 \times 3 \times 4 = 60$

Case 2. Blue - Green - Black : $5 \times 4 \times 3 = 60$

Case 3. Blue - Black - Black : $5 \times 3 \times 2 = 30$

Case 4. Blue - Green - Green : $5 \times 4 \times 3 = 60$

Case 5. Blue - Blue - Black : $5 \times 4 \times 3 = 60$

Case 6. Blue - Black - Blue : $5 \times 3 \times 4 = 60$

Case 7. Blue - Blue - Green : $5 \times 4 \times 4 = 80$

Case 8 Blue - Green - Blue : $5 \times 4 \times 4 = 80$

Case 9. Blue - Blue - Blue : $5 \times 4 \times 3 = 60$

Hence, the total possible outcomes are 550.

14 How many odd numbers with middle digit 5 and no digit repeated are there between 20000 and 69999?

14.6 Combination: Another Tool for Overcounts

Sometimes, we overcount in order to solve harder questions. For instance, check out the following question.

15 In how many ways can Bob choose three of the six colors of Original Skittles to eat? (Make sure he eats one Skittle per flavor.)

This tells us the usage of "division." The purpose of division is to eliminate the overcounts. This result is the number of combinations of k objects from a set of n objects. It is denoted by $\binom{n}{k}$, or sometimes by $_nC_k$.

16 A set of points is chosen on the circumference of a circle so that the number of different triangles with all three vertices among the points is equal to the number of pentagons with all five vertices in the set. How many points are selected in the beginning?

14.7 Permutation vs. Combination

It is easy to get confused between combinations and permutations. The thing to remember is that with permutations, order matters. In general rearrangements, the overcounting concept is still useful. For example, suppose we want to find in how many ways the word NAGINI can be rearranged. We might immediately think the answer is 6!, since there are 6 letters in the word. The problem is that we are overcounting. The two I's are distinct; call them I_1 and I_2. We counted, for example, $NAGI_1NI_2$ and $NAGI_2NI_1$, although they are actually the same! Though distinct, the I's are indistinguishable. Likewise, the N's are indistinguishable.

Among the 6! arrangements, I's and N's been written 2! times by ordering the two letters in 2! ways. Dividing out by all these repetitions, since we only want one copy of each arrangement in the end, the final answer is $\dfrac{6!}{2!2!}$.

17 In how many ways can the word APPLEBERRY be rearranged?

Combination is exactly the same concept of filling the positions in a line with indistiguishables. In fact, if you feel comfortable with counting problems, a good exercise is to do problems in more than one way. This helps you stay flexible and not get locked into one mode of solving.

How many ways can *BANANA* be arranged?

Solution
Let's think of *A*'s. There are three *A*'s. Out of six places, we need to choose three spots to fill up with *A*'s. Hence, $\binom{6}{3}$ ways are there to fill up *A*'s.

Then, out of three remaining spots, we will choose two spots to be filled by *N*'s. There are $\binom{3}{2}$ ways are there to fill up three spots with two *N*'s.

Lastly, out of one remaining spot, we will choose one spot to be filled by one *B*, which is one way of doing so.

Due to the ongoing process of counting, we multiply $\binom{6}{3}$, $\binom{3}{2}$ and $\binom{1}{1}$, which is exactly equal to $\dfrac{6!}{3!2!}$.

18 Given *ABCDEFAAB*, find the total number of distinct arrangements of the word.

1 What is the total number of digits used when the first 1,000 positive even integers are written?

2 Bob counts up from 1 to 9, and then immediately counts down again to 1, and then back up to 9, and so on, alternately counting up and down

$$(1,2,3,4,5,6,7,8,9,8,7,6,5,4,3,2,1,2,3,4,\dots).$$

What is the 500th integer in her list?

3 How many different rectangles with sides parallel to the grid can be formed by connecting four of the dots in a 4×4 square array of dots, as in the figure below?

```
· · · ·
· · · ·
· · · ·
· · · ·
```

(Two rectangles are different if they do not share all four vertices.)

4 In how many ways can 8 people be seated in a row of chairs?

5 In how many ways can 5 people be seated in a round table?

6 In how many ways can 8 people be seated in a row of chairs if three of the people, Bob, Sarah, and Charlie refuse to sit in three consecutive seats?

1 3448

2 4

3 36

4 $8!(=40,320)$

5 $\dfrac{5!}{5}(=24)$

6 $_6P_3 \times 5!(=14,400)$

Solution Manual

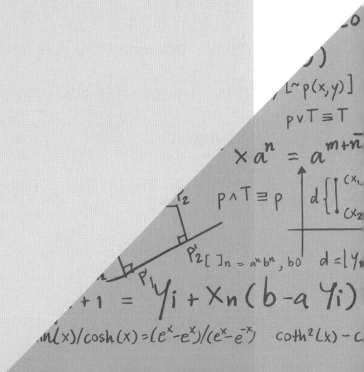

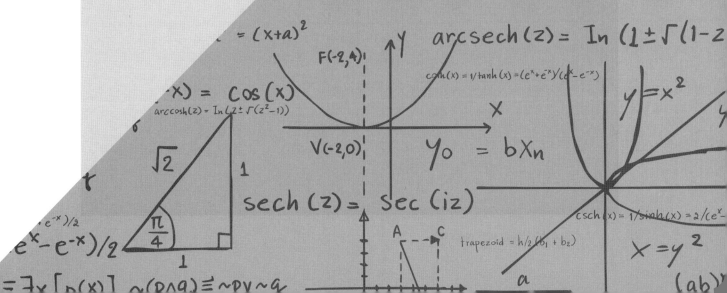

1 We can set up the equation $m = n + 1$ where m is the number of pints of milk of nth cow, and n is the nth cow. Hence, 100th cow gives 101 pints of milk.

2 $1 + 3 + 5 + \cdots + 79 = 1600$.

3 $0 \cdot 1 \cdot 2 \cdots 9 = 0$.

4 $20 \cdot 17 - 10(4 \cdot 6) = 100$ square inches.

5 $(115 + 35) \cdot 205 = 150 \cdot 205 = 150(200 + 5) = 30000 + 750 = 30750$.

6 (a) 468000   (b) 46800

7 $12(45 + 32 + 23) = 12(100) = 1200$.

8 9

9 (a) 13   (b) 9

10 There are 26 times Adam and Bob have the same number written on the board.

11 36

12 Since the difference is positive, the number on the leftside must be bigger than that on the rightside. Hence, the leftmost number is $9 + (-15) = -6$.

13 0 has no reciprocal. Since x is nonzero,
$$x(-\frac{1}{x}) = x(-1)(\frac{1}{x}) = (-1) \cdot x \cdot \frac{1}{x} = -1 \cdot 1 = -1 \text{ and}$$
$$(-x)(\frac{1}{x}) = (-1)(x)(\frac{1}{x}) = -1, \text{ so } (-1) - (-1) = -1 + 1 = 0.$$

14 $\dfrac{1}{-\frac{1}{3} \cdot -\frac{1}{2}} + \dfrac{1}{\frac{1}{4} \cdot -\frac{1}{8}} = \dfrac{1}{\frac{1}{6}} + \dfrac{1}{-\frac{1}{32}} = 6 + (-32) = -26$.

▶ Solution for Topic 2 Questions

1

(a) $(4 \cdot 5)^2 = (20)^2 = 400$

(b) $4^2 \cdot 5^2 = 16 \cdot 25 = 400$

(c) $(ab)^2 = (ab)(ab) = a(ba)b = a(ab)b = (aa)(bb) = a^2 b^2$

2

(a) $(512 \div 16)^2 = 1024$

(b) $512^2 \div 16^2 = 262144 \div 256 = 1024$

(c) $\left(\dfrac{a}{b}\right)^2 = \left(\dfrac{a}{b}\right)\left(\dfrac{a}{b}\right) = \dfrac{(a \cdot a)}{(b \cdot b)} = \dfrac{a^2}{b^2}$

3

(a) $8 + 6(3 - 8)^2 = 8 + 150 = 158$

(b) $5(3 + 4 \cdot 2) - 6^2 = 55 - 36 = 19$

(c) $92 - 45 \div (3 \cdot 5) - 5^2 = 92 - 3 - 25 = 64$

(d) $8(6^2 - 3(11)) \div 8 + 3 = 3 + 3 = 6$

4 13 perfect squares.

5 $3^2 + 6^2 + 9^2 + \cdot + 75^2 = 9(5525) = 49725$

6 $(-1)^{16} + 1^{16} = 1 + 1 = 2$

7 $3^{200} + 3^{200} + 3^{200} = 3 \cdot 3^{200} = 3^{201}$

8

(a) $(2^4)^3 = (16)^3 = 16 \cdot 16 \cdot 16 = 4096$

(b) $2^{4 \cdot 3} = 2^{12} = 2^{10} \cdot 2^2 = 1024 \cdot 4 = 4096$

(c) $\underbrace{a^m \cdot a^m \cdot a^m \cdots a^m}_{n \text{ copies of } a^m} = a^{mn}.$

9 $(2^8)^{10000} < (5^4)^{10000} < (11^3)^{10000}$, so $2^{80000} < 5^{40000} < 11^{30000}$

10 $\underbrace{1+1+1+1+\cdots+1}_{100 \text{ copies of } 1} = 100$

11

(a) $4 \cdot 2^4 = 2^2 \cdot 2^4 = 2^6 = 64$

(b) $2^5(1+2^1+2^2) \div 2^3 = 2^2(1+2+2^2) = 4(1+2+4) = 4 \cdot 7 = 28$

(c) $3 \cdot 3^4 \cdot 2^3 - 2 \cdot 2^3 \cdot 3^4 = 648$

(d) $\left(\dfrac{88888}{22222}\right)^4 = 4^4 = (2^2)^4 = 2^8 = 256$

(e) $4^4(1+4+6+4+1) = 4^4 \cdot 4^2 = 4^6 = 2^{12} = 4096$

(f) $4^{27} \div 4^9 = 4^{18} = 2^{36}$

12 $P = 0^{123456789} = 0$ and $Q = 0^{123456789}$, so
$(1+2+3+4)^{P+Q} = (1+2+3+4)^0 = (10)^0 = 1.$

13 $6^0 \cdot x^{9876} + 6 \cdot x^{9876} = 1 \cdot x^{9876} + 6 \cdot x^{9876} = (1+6)x^{9876} = 7x^{9876}.$

14

(a) $1^{-5} = \dfrac{1}{1^5} = \dfrac{1}{1} = 1.$

(b) $10^{-4} = \dfrac{1}{10^4} = \dfrac{1}{10000}.$

(c) $2^{-3} = \dfrac{1}{2^3} = \dfrac{1}{8}.$

(d) $56 \cdot 2^{-3} = 56 \cdot \dfrac{1}{2^3} = 56 \cdot \dfrac{1}{8} = 7.$

(e) $56 \div 2^{-3} = 56 \cdot 8 = 448.$

(f) $\left(\dfrac{1}{16}\right)^{-(-2)} = \left(\dfrac{1}{16}\right)^2 = \dfrac{1}{16^2} = \dfrac{1}{2^8} = \dfrac{1}{256}.$

15

(a) $3^{-1} \cdot 3^{-2} = \dfrac{1}{3} \cdot \dfrac{1}{9} = \dfrac{1}{27}$.

(b) $3^{15} \cdot 3^{-5} \cdot 3^{-4} \cdot 3^{-3} = 3^{15-5-4-3} = 3^3 = 27$.

16

(a) $\dfrac{1}{2^{-3}} = 8$

(b) $\dfrac{1}{5^{-2}} = 25$

(c) $\dfrac{1}{a^{-n}} = \dfrac{1}{1/a^n} = a^n$.

1 Yes, 192 is a multiple of 12.

2 Yes, $x - 10$ is a multiple of 5.

3 No, $x + 13$ is not a multiple of 3.

4 $7^3 = 343$ is a perfect cube and a multiple of 7 between 300 and 400.

5 $17 \cdot 58 = 986$ is the largest three-digit multiple of 17.

6 There are 143 multiples of 7 between 7 and 1001, inclusive.

7 363637 is not divisible by 3.

8 2 is the only value of X.

9 $Y = 4$.

10 $Z = 4$ is the only possible value of Z.

11 There is only one prime 97 between 90 and 100.

12 Bob has 31 students in his class because 31 is the only prime between 30 and 36, inclusive.

13 173 is the largest prime less than 200, none of whose digits are composite.

14

(a) 3   (b) 3   (c) 2   (d) 5   (e) 5   (f) 7

15

(a) prime   (b) $3^3 \cdot 11$ (c) $17 \cdot 19$ (d) prime   (e) $7 \cdot 79$ (f) prime

16 There is only one multiple of 3, i.e., 3.

17

(a) prime   (b) $17 \cdot 23$ (c) prime   (d) prime

18

(a) $30 = 2 \cdot 3 \cdot 5$

(b) $60 = 2^2 \cdot 3 \cdot 5$

(c) $252 = 2^2 \cdot 3^2 \cdot 7$

(d) $243 = 3^5$

19

(a) $\text{lcm}[96, 144] = 2^5 \cdot 3^2 = 288$

(b) $\text{lcm}[24, 36, 42] = 2^3 \cdot 3^2 \cdot 7 = 504$

20

(a) $\gcd(72, 240, 288) = 2^3 \cdot 3 = 24$

(b) $\gcd(14, 42) = 2 \cdot 7 = 14$

1

(a) There are $\left\lfloor \dfrac{40}{7} \right\rfloor = 5$ values of n, i.e., $7, 14, 21, 28, 35$.

(b) There are $\left\lfloor \dfrac{100}{8} \right\rfloor = 12$ values of n, i.e., $8, 16, \cdots, 96$.

2 $\dfrac{53}{5}$ is between 10 and 11.

3 Computing it directly, we get 13.

4 $2xy - \dfrac{x}{2y} = -\dfrac{693}{10}$ when $x = -7$ and $y = 5$.

5

(a) $\dfrac{55}{42}$

(b) -10

(c) 70

6 $\dfrac{27 \cdot 25 \cdot 22 \cdot 20}{3 \cdot 4 \cdot 5 \cdot 11} = 450$

7 Canceling out common numbers, we get $\dfrac{12}{5}$.

8 $\dfrac{4}{5} \cdot \dfrac{5}{12} \cdot 123456 = 41152$.

9 $\dfrac{x}{y} \cdot \dfrac{y}{z} = \dfrac{7}{15} \cdot \dfrac{12}{19} = \dfrac{28}{95}$.

10 $\dfrac{35}{72}$.

11 It has the same effect as multiplying the original number by $\dfrac{5}{4}$.

12

(a) $\dfrac{9}{4}$

(b) 243

(c) $\dfrac{1}{1024}$

13 1

14

(a) $\dfrac{7}{12}$

(b) $\dfrac{3}{56}$

(c) $-\dfrac{1}{5}$

(d) $-\dfrac{3}{4}$

15 $\dfrac{127}{128} + \dfrac{33}{32}$ is closest to 2.

1 Let x be the number of pieces of gum in a pack, i.e., 1 pack $= x$ pieces of gum. Then, 5 packs include $5x$ pieces of gum and 6 packs include $6x$ pieces of gum. Hence, $5x + 6x = 11x$ is the total number of pieces of gum for 11 packs.

2 $3x + 7 = 4x + 5$ implies $x = 2$ (candies).

3 $\dfrac{17k+4}{7} - \dfrac{3-5k}{5} = \dfrac{5(17k+4) - 7(3-5k)}{35} = \dfrac{120k-1}{35}.$

4 Since the initial value is $29 - 2x$ and the common difference is $(2x+1) - (3-x)$ per day, we get

$$(29 - 2x) + 3(2x + 1 - (3 - x)) = (29 - 2x) + 3(3x - 2) = 29 - 2x + 9x - 6 = 23 + 7x$$

5 By isolating the variable, we get $N = 28$.

6 $\dfrac{1}{n} = \dfrac{1}{2^5} = \dfrac{1}{32}$ implies $n = 32$.

7 $5p - 3p + 12 = 6p + 6$ implies $p = \dfrac{3}{2}$.

8 Solving $\dfrac{7}{18} = \dfrac{p}{72}$, we get $p = 28$. Also, $\dfrac{7}{18} = \dfrac{28+q}{108}$ implies $q = 14$. Lastly, $\dfrac{r-28}{126} = \dfrac{7}{18}$ implies $r = 77$.

9 Rearranging the terms to isolate the variable x, we get $x = -\dfrac{2}{3}$.

10 Let x, $2x$, and $4x$ be the three positive integers. Then, the difference between any two numbers is either x, $2x$, or $3x$. Since one of the expressions must equal 23, x must be 23. Hence, the largest integer of the three numbers must be $4(23) = 92$.

11 Let Harry's current age be x. Then, Joshua's current age is $x - 15$. Hence, $(x+6) = 2(x - 15 + 6)$ implies $x = 24$. Harry must be 24 years of age.

12 Let x be the number of nickels. Then, there are $3x$ coins in total. According to the question, we get $0.05x + 0.1x + 0.25x = 9.2$, implying that $x = 23$. Hence, there are $3(23) = 69$ coins in total.

13 $x(x + \frac{1}{x}) = 2$. Hence, $x^2 + 1 = 2$, so $x^2 = 1$. Since x is assumed to be positive, we get $x = 1$.

14 Let x be the number of cars. Then, the number of motocycles is $27 - x$. Solving $4x + 2(27 - x) = 72$, we get $x = 9$. There are 9 cars and 18 motocycles.

15 Solving $n < \frac{22}{4}$ and $-\frac{19}{7} < n$ at the same time, we get $n = -2, -1, 0, 1, 2, 3, 4, 5$. There are 8 values of n, in total.

16 $\frac{x}{6} + 7 \geq 20$ implies $x \geq 78$. Hence, the least possible number of chocolate muffins in the batch is 78.

17 Due to the assumption, a customer 'saves' money if $150 + 29x \leq 44x$. Hence, $10 \leq x$ implies that 10 cups is the minimum number of cups the customer should buy to 'save' money.

18 Let y be the number of chocolate roll-cakes and x be the number of cheese roll-cakes. Then,

$$2 + \frac{2}{3}y \leq x \leq 2y$$

Since $2 + \frac{2}{3}y \leq 2y$ implies $y = 2, 3, 4, \cdots$, substitute $y = 2$ to the inequality. Then, $2 + \frac{4}{3} \leq x \leq 4$ implies that $x = 4$. Hence, $x + y = 2 + 4 = 6$ is the least number of roll-cakes that Bob can bring.

1

(a)

$$\begin{aligned}
3.1 + 2.4 &= 3 \cdot 10^0 + 1 \cdot 10^{-1} + 2 \cdot 10^0 + 4 \cdot 10^{-1} \\
&= (3+2) \cdot 10^0 + (1+4) \cdot 10^{-1} \\
&= 5 + 5 \cdot 10^{-1} \\
&= 5.5
\end{aligned}$$

(b)

$$\begin{aligned}
12.3 + 21.45 &= 1 \cdot 10^1 + 2 \cdot 10^0 + 3 \cdot 10^{-1} + 2 \cdot 10^1 + 1 \cdot 10^0 + 4 \cdot 10^{-1} + 5 \cdot 10^{-2} \\
&= 3 \cdot 10^1 + 3 \cdot 10^0 + 7 \cdot 10^{-1} + 5 \cdot 10^{-2} \\
&= 33.75
\end{aligned}$$

(c)

$$\begin{aligned}
3 - 2.31 &= 3 \cdot 10^0 - (2 \cdot 10^0 + 3 \cdot 10^{-1} + 1 \cdot 10^{-2}) \\
&= 1 \cdot 10^0 - 3 \cdot 10^{-1} - 1 \cdot 10^{-2} \\
&= 10 \cdot 10^{-1} - 3 \cdot 10^{-1} - 1 \cdot 10^{-2} \\
&= 7 \cdot 10^{-1} - 1 \cdot 10^{-2} \\
&= 70 \cdot 10^{-2} - 1 \cdot 10^{-2} \\
&= 69 \cdot 10^{-2} \\
&= 6 \cdot 10^{-1} + 9 \cdot 10^{-2} \\
&= 0.69
\end{aligned}$$

2

(a)

$$\begin{aligned}
3.56 \div 10 &= (3 + 5 \cdot 10^{-1} + 6 \cdot 10^{-2}) \div 10 \\
&= (3 + 5 \cdot 10^{-1} + 6 \cdot 10^{-2}) \times 10^{-1} \\
&= 3 \cdot 10^{-1} + 5 \cdot 10^{-2} + 6 \cdot 10^{-3} \\
&= 0.356
\end{aligned}$$

(b)

$$12.34 \cdot 100 = (1 \cdot 10^1 + 2 \cdot 10^0 + 3 \cdot 10^{-1} + 4 \cdot 10^{-2}) \times 10^2$$
$$= 1 \cdot 10^3 \cdot 2 \cdot 10^2 \cdot 3 \cdot 10^1 + 4 \cdot 10^0$$
$$= 1234$$

(c)

$$0.00023 \div 100 = (2 \cdot 10^{-4} + 3 \cdot 10^{-5}) \cdot 10^{-2}$$
$$= 2 \cdot 10^{-6} + 3 \cdot 10^{-7}$$
$$= 0.0000023$$

3

(a)

$$2.3 \cdot 4 = (2 \cdot 10^0 + 3 \cdot 10^{-1}) \cdot 4$$
$$= 8 \cdot 10^0 + 12 \cdot 10^{-1}$$
$$= 8 \cdot 10^0 + (10 + 2) \cdot 10^{-1}$$
$$= 9 \cdot 10^0 + 2 \cdot 10^{-1}$$
$$= 9.2$$

(b)

$$1.3 \cdot 2.3 = (1 + 3 \cdot 10^{-1})(2 + 3 \cdot 10^{-1})$$
$$= (2 + 3 \cdot 10^{-1} + 6 \cdot 10^{-1} + 9 \cdot 10^{-2}$$
$$= 2 + 9 \cdot 10^{-1} + 9 \cdot 10^{-2}$$
$$= 2.99$$

(c)

$$0.012 \cdot 0.001 = (1 \cdot 10^{-2} + 2 \cdot 10^{-3}) \cdot 10^{-3}$$
$$= 1 \cdot 10^{-5} + 2 \cdot 10^{-6}$$
$$= 0.000012$$

4

(a)

$$\begin{aligned}
(0.12)^2 &= (1 \cdot 10^{-1} + 2 \cdot 10^{-2})^2 \\
&= (1 \cdot 10^{-1} + 2 \cdot 10^{-2})(1 \cdot 10^{-1} + 2 \cdot 10^{-2}) \\
&= 1 \cdot 10^{-2} + 2 \cdot 10^{-3} + 2 \cdot 10^{-3} + 4 \cdot 10^{-4} \\
&= 1 \cdot 10^{-2} + 4 \cdot 10^{-3} + 4 \cdot 10^{-4} \\
&= 0.0144
\end{aligned}$$

(b)

$$\begin{aligned}
1 \div 0.04 &= 1 \times (4 \times 10^{-2})^{-1} \\
&= 1 \times 4^{-1} \times 10^2 \\
&= \frac{100}{4} \\
&= 25
\end{aligned}$$

5

(a) $43_{(8)}$ (b) $432_{(8)}$ (c) $7654_{(8)}$ (d) $500300_{(8)}$

6 $888_{(9)} = 8 \cdot 9^2 + 8 \cdot 9^1 + 8 \cdot 9^0 = 728$ is the largest three-digit number in base 9.

7

(a) 900

(b) -2000

(c) 12.3

(d) 0.009

8 $2.85 \leq x < 2.95$

9 The reciprocal of $\dfrac{21}{10}$ is $\dfrac{10}{21}$.

10

(a) $\dfrac{3}{8} = \dfrac{3}{2^3} = \dfrac{3 \cdot 5^3}{2^3 \cdot 5^3} = \dfrac{375}{1000} = 0.375$

(b) $\dfrac{2}{5} = \dfrac{2 \cdot 2}{5 \cdot 2} = \dfrac{4}{10} = 0.4$

(c) $\dfrac{19}{100} = 0.19$

11

(a) $\dfrac{5}{7} = 0.\overline{714285}$

(b) $\dfrac{2}{9} = 0.\overline{2}$

(c) $\dfrac{11}{13} = 0.\overline{846153}$

12 Since $\dfrac{2}{7} = 0.\overline{285714}$, the bundle of size 6 repeats 8 times with the remainder of 2. Hence, the 50th digit is the second number in the 9th bundle of 285714. Hence, 8 is the correct digit.

13 Let $x = 0.\overline{9}$. Then, $10x = 9.\overline{9}$. Hence, $10x - x = 9.\overline{9} - 0.\overline{9} = 9$, so $x = 1$.

1 $f:m=4:5$, so $f:25=4:5$. Hence, $f=20$. Therefore, $f+m=20+25=45$ dogs are at a facility.

2 Let x be the short piece, then the long piece must be $10-x$. Since the ratio between the two satisfies $2:3$, we get $x:10-x=2:3$. Hence, $3x=20-2x$. Therefore, $5x=20$, so $x=4$ (inches).

3 Since $4:5=396:x$, we get $m=495$ students. Hence, the incoming class size is $891(=396+495)$. Hence, $2:11=p:891$, so $p=162$ professors.

4 There are 6 losses and 4 wins in the original games. Since there are additional 5 losses and wins, the new ratio of losses to wins is 11 to 9.

5 Let x be the number of black marbles and y the number of blue marbles in a jar. Then, $x:y=2:5$. Since $x+y=245$, there are 70 black marbles and 175 blue marbles. Now, let b be the number of newly added black marbles. Then, $70+x:175=3:7$. Hence, $7(70+x)=3(175)$, so $x=5$ black marbles.

6

(a) $1:2:3$ (b) $3:4:1:2$ (c) $3:4:2$

7 $\$1,000,000 \times \dfrac{5}{5+2+1} = \$625,000$.

8 Let x be the amount of butter, y the amount of flour, and z the amount of milk. Then, $x:y:4:z=1:6:2:1$. Therefore, there should be $x=2$ cups of butter, $y=12$ cups of flour, $z=2$ cups of milk.

9 $\$338 \times \dfrac{6}{6+4+3} = \$156$.

10 4 inch $\times \dfrac{5\text{ km}}{1\text{ inch}} = 20$ km.

11 The scaled-up photo must be 4 cm wide × 8 cm tall.

12 $48 \text{ inches} \times \dfrac{1 \text{ foot}}{12 \text{ inches}} \times \dfrac{1 \text{ yard}}{3 \text{ feet}} = \dfrac{4}{3} \text{ yard}.$

13 $4 \text{ gallons} \times \dfrac{16 \text{ cups}}{1 \text{ gallon}} \times \dfrac{8 \text{ ounce}}{1 \text{ cup}} \times \dfrac{1 \text{ tablespoon}}{\frac{1}{2} \text{ ounce}} = 1024 \text{ tablespoons}.$

14 $10 \text{ quarts} \times \dfrac{1 \text{ gallon}}{4 \text{ quarts}} \times \dfrac{8 \text{ pounds}}{1 \text{ gallon}} = 20 \text{ pounds}.$

15

(a) 150 km   (b) 75 km/h   (c) 4 hours

16 Since Bob jogs for $\dfrac{3}{4}$ hours at 2 km/h, he would have run $\dfrac{3}{2}$ km. Since Bob runs for $\dfrac{1}{2}$ hours at 6 km/h, he would have run additional 3 km. Hence, Bob would have gone 4.5 km for one hour and 15 minutes. Therefore, the average speed is $\dfrac{18}{5}(= 3.6)$ km/h.

17 On his way to the office, Bob drove for $\dfrac{1}{4}$ hour. However, his return trip took 1 hour. In total, it takes about $\dfrac{5}{4}$ hours to drive 40 km. Hence, the average speed for the entire trip is 32 km/h.

18 $2000 \text{ word} \times \dfrac{1 \text{ min}}{50 \text{ words}} = 40 \text{ minutes}.$

19
$5100 \text{ gallons} \times \dfrac{1 \text{ second}}{1 \text{ gallon}} \times \dfrac{1 \text{ minute}}{60 \text{ seconds}} \times \dfrac{1 \text{ hour}}{60 \text{ minutes}} = \dfrac{85}{60} \text{ hours} = 1\dfrac{5}{12} \text{ hours}.$

20 $60 \text{ minutes} \times \dfrac{150 \text{ words}}{1 \text{ minutes}} \times \dfrac{1 \text{ page}}{500 \text{ words}} = 18 \text{ pages}.$

1

(a) 0.34   (b) 1.35   (c) −4.2

2

(a) 23%   (b) −412.5%   (c) $66.\overline{6}\%$ (d) −12500%

3 Since $\dfrac{x}{100} \times 36 = 9$, we get $x = 25(\%)$.

4 Since $15 = \dfrac{10}{100} \times x$, we get $x = 15 \times 10 = 150$.

5 $\dfrac{35}{100} \times 40 = 14$.

6 Since the tax amount is $\$800 \times \dfrac{15}{100} = \$120$, the actual price Bob has to pay is $800 + 120 = 920$ dollars.

7 In total, there are $800 (= 500 + 300)$ students. The percentage of boys is $100 \times \dfrac{300}{800} = 37.5(\%)$.

8 Let x be the number of students in the class. Since 20% of the students failed, we get $x \times \dfrac{20}{100} = 5$. Therefore, $x = 25($ students$)$.

9 $\dfrac{25 - 20}{20} \times 100 = 25\%$.

10 $\dfrac{25 - 20}{25} \times 100 = 20\%$.

(a) $\dfrac{13-10}{10} \times 100 = 30\%$ increase.

(b) $\dfrac{40-33}{40} \times 100 = 17.5\%$ decrease.

(c) $\dfrac{20-14}{20} \times 100 = 30\%$ decrease.

(d) $\dfrac{13-5}{13} \times 100 = 61\dfrac{7}{13}\%$ decrease.

(e) $\dfrac{4.2-4}{4} \times 100 = 5\%$ increase.

(f) $\dfrac{3.5-3}{3} \times 100 = 16\dfrac{2}{3}\%$ increase.

1

(a)

$$x^2 = 16$$
$$\sqrt{x^2} = \sqrt{16}$$
$$|x| = 4$$
$$x = \pm 4$$

(b) Normally, we say there is no real x satisfying the equation. However, we can find complex solutions, using $i = \sqrt{-1}$.

$$x^2 = -25$$
$$\sqrt{x^2} = \sqrt{-25}$$
$$|x| = 5i$$
$$x = \pm 5i$$

(c)

$$x^2 = 1024$$
$$\sqrt{x^2} = \sqrt{1024}$$
$$|x| = 32$$
$$x = \pm 32$$

2 If the equation involves square roots, there should be one solution.

(a) $x = \sqrt{9} \implies x = 3$.

(b) $x = -\sqrt{0} \implies x = -0 \implies x = 0$.

(c) If we limit our attention to the set of real numbers, then there is no such x satisfying the equation. On the other hand, if we extend it to the set of complex numbers, then $x = \sqrt{-4} = \sqrt{4}\sqrt{-1} = 2i$.

3

(a) $\sqrt{225} = \sqrt{15^2} = \sqrt{15}\sqrt{15} = 15$.

(b) $\sqrt{144} = \sqrt{12^2} = \sqrt{12}\sqrt{12} = 12$.

4

(a) $\sqrt{14^2} = |14| = 14$.

(b) $\sqrt{123456^2} = |123456| = 123456$.

(c) $\sqrt{(-121)^2} = |-121| = 121$.

(d) $\sqrt{5^6} = |5^3| = 5^3 = 125$.

5

(a) $\sqrt{(5 \cdot 35 \cdot 7)^2} = |5 \cdot 35 \cdot 7| = 35^2 = 1225$.

(b) $\sqrt{64 \cdot 49} = \sqrt{64}\sqrt{49} = 8 \cdot 7 = 56$.

(c) $\sqrt{360000} = \sqrt{36}\sqrt{10000} = 6 \cdot 100 = 600$.

6 $\sqrt{8 \cdot 8^7} = \sqrt{8^8} = 8^4 = (2^3)^4 = 2^{12} = 4096$.

7 Since $1.5 = \sqrt{(1.5)^2} = \sqrt{2.25}$, $\sqrt{3}$ is greater than 1.5.

8 $\sqrt{15} < \sqrt{16}(=4) < \cdots < \sqrt{121} < \cdots < \sqrt{142} < \sqrt{144}(=12)$, so $4, 5, 6, \cdots, 11$ are the integers. There are 8 integers between $\sqrt{15}$ and $\sqrt{142}$.

9 $6\sqrt{11} = \sqrt{36 \times 11} > \sqrt{25 \times 13} = 5\sqrt{13}$.

10 $\sqrt{16}\sqrt{36} = \sqrt{16 \times 36} = \sqrt{n}$, so $n = 16 \times 36 = 576$.

11 $(\sqrt{7}\sqrt{5})^2 = (\sqrt{7}\sqrt{5})(\sqrt{7}\sqrt{5}) = (\sqrt{7}\sqrt{7})(\sqrt{5}\sqrt{5}) = 7 \times 5 = 35$.

12 $\sqrt{5}\sqrt{7} = \sqrt{5 \times 7} = \sqrt{5 \times n}$ implies $n = 7$.

13

(a) $\sqrt{3}\sqrt{12} = \sqrt{36} = 6.$

(b) $\sqrt{18}\sqrt{32} = 3\sqrt{2} \times 4\sqrt{2} = 12 \times 2 = 24.$

(c) $(3\sqrt{3})(5\sqrt{27}) = 15\sqrt{3} \times 3\sqrt{3} = 45 \times 3 = 135.$

14

(a) $\sqrt{\dfrac{36}{9}} = \sqrt{4} = 2.$

(b) $\sqrt{\dfrac{27}{192}} = \sqrt{\dfrac{9}{64}} = \dfrac{\sqrt{9}}{\sqrt{64}} = \dfrac{3}{8}.$

(c) $\sqrt{11\dfrac{1}{9}} = \sqrt{\dfrac{100}{9}} = \dfrac{\sqrt{100}}{\sqrt{9}} = \dfrac{10}{3}.$

(d) $\dfrac{\sqrt{96}}{\sqrt{6}} = \sqrt{\dfrac{96}{6}} = \sqrt{16} = 4.$

(e) $\dfrac{\sqrt{112}}{\sqrt{28}} = \dfrac{\sqrt{7}\sqrt{16}}{\sqrt{7}\sqrt{4}} = \dfrac{\sqrt{16}}{\sqrt{4}} = \dfrac{4}{2} = 2.$

1 Since $\angle AOB$ and $\angle BOC$ are adjacent and complementary, $3x - 10 + x = 90$. $4x = 100$ implies $x = 25$.

2 $15° + 90° = 105°$.

3 The minute hand makes $120°$ from 4-hour to 8-hour position. On the other hand, the hour hand makes $10°$ from 4-hour position. Hence, the sum of the two angles $10° + 120° = 130°$ is the angle formed by the hands of a clock at $3:40$.

4

Vertical Angles	(1,3), (2,4), (5,7), (6,8)
Alternate Interior Angles	(4,6), (3,5)
Alternate Exterior Angles	(1,7), (2,8)
Consecutive Interior Angles	(4,5), (3,6)
Consecutive Exterior Angles	(1,8), (2,7)
Corresponding Angles	(1,5), (2,6), (3,7), (4,8)

5 Let x be the angle. Then,

$$180 - x = 5(90 - x)$$
$$180 - x = 450 - 5x$$
$$4x = 450 - 180$$
$$4x = 270$$
$$x = \frac{270}{4} (= 67.5°)$$

6 Let $m\angle MBQ = a$. Then, $m\angle MBP = a$. Hence, $m\angle QMP = 2a$. Since BC trisect $\angle ABC$, we get $m\angle ABQ = 4a$. Therefore, the ratio of the mreasure of $\angle MBQ$ to the measure of $\angle ABQ$ is $1:4$.

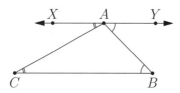

Since $\overline{XY}$ is parallel to $\overline{BC}$, $\angle XAC \cong \angle ACB$ and $\angle YAB \cong \angle ABC$. Since $\angle XAC, \angle CAB, \angle YAB$ are supplementary, the sum of the interior angles is always $180°$.

8

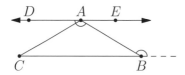

First, $\overline{DE}$ is parallel to $\overline{BC}$, so $\angle BAD \cong \angle B$. Also, $\angle DAC \cong \angle BCA$ because they form an alternate interior angle pair. Hence, $m\angle C + m\angle A = m\angle B$ where $\angle B$ is the exterior angle of B.

9 The measure of $\angle BDC$ is $51° (= 180° - 129°)$.

10 Since $2x = 112°$, we get $x = 56°$. Therefore, $m\angle PQR = 124° (= 180° - 56°)$.

11 The measure of $\angle BQC$ is $88° (= 180° - 92°)$.

12 Since the exterior angle is $10°$, and there are n sides in regular polygon to form $n \times 10° = 360°$, we get $n = 36$.

13 $m\angle BAD = 180° - \dfrac{360°}{7} = \dfrac{900°}{7}$.

1 Since $CB = AB$, and $2CD = AC$, we get $10 + 10 + 5 + 5 + 5 = 35$ units.

2 Let x be the base length. Then, the leg length is $2x$, each, so $2x + 2x + x = 25$, i.e., $x = 5$.

3 There are two possible rectangles to be drawn. First, we can cut 8-foot length into four congruent parts, so there are four congruent rectangles of 4-foot by 2-foot. On the other hand, we may cut 4-foot width into four congruent parts, so there are four congruent rectangles of 1-foot by 8 foot. The greatest possible perimeter of a single piece is 18 feet, whereas the smallest possible perimeter of a single piece is 12 feet. Hence, the sum of the two values is 30 feet.

4 Let x be the side length of the smallest square. Then, the largest square that is not shaded has the side length of $4x$. Since $ABCD$ is a rectangle, we get $AB = CD$, so $4x + x = 30$. Hence, $x = 6$. Therefore, the largest square that is not shaded has the side length of $24(= 4x)$.

5

(a) 22

(b) 16

(c) 22

6 Let the width of the rectangle be s and the length of it be h. Then, $\triangle ADQ$ has the area of $\frac{1}{2} \times h \times \frac{s}{2}$. Likewise, $\triangle PQC$ has the area of $\frac{1}{2} \times \frac{s}{2} \times \frac{h}{2}$. Also, $\triangle ABP$ has the area of $\frac{1}{2} \times s \times \frac{h}{2}$. Summing up the expressions, we get $\frac{5}{8}hs$ where $hs = 64$, as given. Hence, the area of the triangle APQ is $64 - \frac{5}{8}(64) = 24$.

7 Rearranging the figures, we get the area of 9 because there are 9 square grids that can be contained inside the parallelogram.

8 Since X is the midpoint of AC, $[\triangle ABX] = [\triangle BXC] = 20$. Also, Y is the midpoint of BX, implying that $[\triangle ABY] = [\triangle AYX] = 10$. Hence, $[\triangle AYX] = 10$.

9

(a) 12 square inches

(b) 13 square inches

(c) 8 square inches

10 $2\pi r = 28\pi$, so $r = 14$ units.

11 We can subtract the right isosceles triangle from the quarter circle to get the half of the bounded region. This region can be computed by

$$\frac{1}{4} \times 10 \times 10 \times \pi - \frac{1}{2} \times 10 \times 10$$

Since there are two of these shapes, we get $2(25\pi - 50) = 50\pi - 100$ square units.

12 Because circles are similar, we will use the similarity ratio. The length ratio between the small circle to the large circle is $40 : 120 (= 1 : 3)$, then the area ratio must be $1^2 : 3^2 = 1 : 9$. Since there are 100 trees in the small circular garden, there must be $9(100) = 900$ trees in the large circular garden.

13 A line and a circle have at most two intersection points. There are three lines and one circle, so there are at most 6 intersection points. On the other hand, three lines can meet at three points at most. Therefore, there are $9(= 6 + 3)$ points of intersection of these figures.

14 The ratio of circumference equals the ratio of length. Since the length ratio is $2 : 3$, the area ratio must be $4 : 9$.

1 Let the ratio of width to length be $3 : 4$. Then, the actual length must be $3k$ and $4k$, for some k, respectively. Therefore, Pythagorean Theorem states that the diagonal must be $5k$ for some k. Since the perimeter is 28, we get $2(3k + 4k) = 14k = 28$, so $k = 2$. Therefore, the diagonal must be $5k = 5(2) = 10$ meters.

2 In $\triangle CDA$, we get $AD = 12$ from $5 : 12 : 13$. Hence, $BD = 9$ from $3 : 4 : 5$ ratio. Therefore, $BC = BD - CD = 9 - 5 = 4$.

3 Since $4^2 = 2^2 + w^2$, we get $w = 2\sqrt{3}$. Hence, the area of the rectangle is $2 \times 2\sqrt{3} = 4\sqrt{3}$.

4 The diagonal is given as 30-inch. Hence, the length must be 24 inches and the height 18 inches. The area of it must be $24 \times 18 = 432$ inch^2.

5 Using the ratio of $1 : 1 : \sqrt{2}$, we get the length of isosceles triangle as $6 : 6 : 6\sqrt{2}$. Therefore, the area must be $\dfrac{1}{2} \times 6 \times 6 = 18$.

6

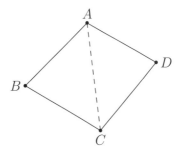

The sum of interior angles of $\triangle ABC$ is $180°$. Likewise, the sum of interior angles of $\triangle ACD$ is $180°$, as well. Hence, the sum of interior angles for a convex quadrilateral is $360°$.

7 Let $x = m\angle D$. Then, $m\angle C = 2x$, $m\angle B = 3x$, and $m\angle A = 6x$. Since the sum of interior angles is $360°$, we get $x + 2x + 3x + 6x = 12x = 360°$. Then, $x = 30°$. Bob feels odd about this because $m\angle A = 180°$, which is a straight angle. Quadrilateral $ABCD$ is no longer quadrilateral. In fact, it turns out to be a triangle.

8 $[ACB] = 20 \times h \times \dfrac{1}{2} = 10h$ where h is the height of the trapezoid. Similarly, $[ABCD] = (20 + 12) \times h \times \dfrac{1}{2} = 16h$. Hence, $\dfrac{[ACB]}{[ABCD]} = \dfrac{10h}{16h} = \dfrac{5}{8}$.

9 Let $PQ = a$. Then, $SR = 2a$. Then, $3a \times h \times \dfrac{1}{2} = 15$, so $ah = 10$. Since $[SQR] = 2a \times h \times \dfrac{1}{2} = ah$, we get $[SQR] = 10$.

10 The area of the parallelogram is $15 \times 4\sqrt{3} = 60\sqrt{3}$, whereas the half of the the parallelogram is $30\sqrt{3}$.

11 $AB = CD$ implies $2x + 4 = 38$, so $x = 17$. Similarly, $AC = BC$ implies $24 = 2y$, so $y = 12$. Hence, the sum of the x and y is $17 + 12 = 29$. (Likewise, the difference of x and y is $17 - 12 = 5$.)

12 The area of the rhombus is 24, whereas the perimeter is 20.

13 The side length of the rhombus is 10, using the special right triangle ratio of $3 : 4 : 5$. Hence, the perimeter is 40.

14

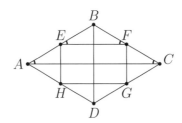

Let $m\angle CAB = x$. Then, $m\angle FEB = x$ by similar triangle properties. Since EH is perpendicular to AC, $m\angle AEH = 90° - x$. Therefore, $m\angle E = 90°$. Using same deduction to F, G, and H, we get $m\angle F = m\angle G = m\angle H = 90°$. Hence, $EFGH$ is a rectangle.

15 Since $xy = 300$ and $2x = 3y$, we get $y = \dfrac{2x}{3}$, so $\dfrac{2x^2}{3} = 300$. Hence, $x^2 = 450$, so $x = 15\sqrt{2}$.

16 Let $ABCD$ has a side length of s. Then, $BDFG$ has the side length of $\sqrt{2}s$. Hence,

$$\frac{[BDFG]}{[ABCD]} = \frac{2s^2}{s^2} = 2$$

17 Let $PA = PB = PC = x$. Then, the Pythagorean theorem states that $x^2 = (6-x)^2 + 3^2$, so $12x = 45$, i.e., $x = 15/4$.

1

1. Faces : DFE, ACB, $ACFD$, $CBEF$, and $ABED$.

2. Edges : $\overline{AD}$, $\overline{EB}$, $\overline{FC}$, $\overline{AB}$, $\overline{BC}$, $\overline{CA}$, $\overline{DE}$, $\overline{FE}$, and $\overline{DF}$.

3. Vertices : A, B, C, D, E, F

2 $\left(\dfrac{3}{2} \times 2 \times \dfrac{1}{2}\right) \times 3 = \dfrac{9}{2}.$

3 $\pi 1^2 \times 5 = 5\pi.$

4 $\left(\dfrac{1}{2} \times (2+4) \times \sqrt{5}\right) \times 3 \times \dfrac{1}{3} = 3\sqrt{5}.$

5 $\pi 2^2 \times 5 \times \dfrac{1}{3} = \dfrac{20\pi}{3}.$

6 $2 \times \left(\dfrac{1}{2} \times 4 \times 3\right) + (3+4+5) \times 5 = 72.$

7 $\pi 3^2 + 4 \times (2\pi(3)) = 42\pi.$

8 $\left(4 \times 2 \times \dfrac{1}{2}\right) \times 2 + \left(\dfrac{\sqrt{67}}{2} \times 1 \times \dfrac{1}{2}\right) \times 2 + (2 \times 1) = 10 + \dfrac{\sqrt{67}}{2}.$

9 According to the formula, $\pi(5)(3) + \pi(3^2) = 24\pi.$

1

The stem-leaf plot is given by

Stem	Leaf
0	1,2,3,4,5,6

On the other hand, the box-whisker plot can be drawn like

2 Let $a, b, 5, 8, 8$ be the distribution of the five integers. Since the mean value is 5,

$$\frac{a+b+5+8+8}{5} = 5$$

where $a+b=4$. Then, $(a,b) = (1,3), (2,2)$, but $(2,2)$ can't be the values of a and b because of the single mode condition. Hence, $a=1$ and $b=3$.

The range is the difference between maximum and minimum, i.e., $8-1=7$. Similarly, the interquartile range is the difference between the upper quartile and the lower quartile, i.e., $8-2=6$.

3 There are $119(=174-56+1)$ integers.

4 There are 216 integers in total.

5 There are $26^4(=456,976)$ four-letter words in total.

6 There are $5 \times 21 \times 21 \times 5(=11,025)$ four-letter words with vowels in each end, and consonants in the middle two places.

7 Bob can distribute skittles to his five students in $6^5(=7,776)$ ways.

8 There are $4 \times 10 \times 1 \times 10 \times 5(=2,000)$ odd numbers with third digit 1, between 30000 and 69999, inclusive.

9 There are $3 \times 2 \times 10 \times 10 \times 9 \times 8(=43,200)$ ways to write a 6-letter word satisfying the given condition.

10 $\dfrac{5!}{4!} \times \dfrac{3!}{2!} \times \dfrac{1!}{0!} = 5 \times 3 = 15.$

11 There are $_{10}P_2 = 10 \times 9 = 90$ possibilities to assign two students in two positions.

12 There are $\dfrac{3!}{3}(=2)$ ways for three students to be seated in a round table.

13 We can do the caseworks. Let c_1 be the first child and c_2 be the second child. Let R, P, O be the colors of the skittles.

Case 1. $(c_1, c_2) = (P, R) : 4 \times 3 = 12$

Case 2. $(c_1, c_2) = (P, O) : 4 \times 5 = 20$

Case 3. $(c_1, c_2) = (R, O) : 3 \times 5 = 15$

Case 4. $(c_1, c_2) = (R, P) : 3 \times 4 = 12$

Case 5. $(c_1, c_2) = (O, P) : 5 \times 4 = 20$

Case 6. $(c_1, c_2) = (O, R) : 5 \times 3 = 15$

Adding all up, we get **94** ways in total.

14 Just as we did in **13**, we do the caseworks.

Case 1. $\boxed{2} \_ \boxed{5} \_ \boxed{3} : 3 \times 7 \times 6$

Case 2. $\boxed{2} \_ \boxed{5} \_ \boxed{7} : 4 \times 7 \times 6$

Case 3. $\boxed{2} \_ \boxed{5} \_ \boxed{9} : 4 \times 7 \times 6$

Case 4. $\boxed{2} \_ \boxed{5} \_ \boxed{1} : 4 \times 7 \times 6$

Adding all up, we get **630** odd numbers satisfying the condition.

15 Bob can choose 2 distinct skittles in $_6C_2 = \dfrac{6 \times 5}{2} = 15$ ways.

16 Given n points, $_nC_3 = {_nC_5}$. Hence, $n = 3 + 5 = 8$ points are on the circle.

17 APPLEBERRY can be arranged in $\dfrac{10!}{2!2!2!}(= 453,600)$ ways.

18 ABCDEFAAB can be arranged in $\dfrac{9!}{3!2!}(= 30,240)$ ways.

The Essential Guide to **Prealgebra**

초판1쇄 발행   2020년 5월 20일
개정1판3쇄 발행   2023년 7월 15일

저자 유하림
발행인 최영민
발행처 헤르몬하우스
주소 경기도 파주시 신촌로 16
전화 031 – 8071 – 0088
팩스 031 – 942 – 8688
전자우편 hermonh@naver.com
출판등록 2015년 3월 27일
등록번호 제406 – 2015 – 31호
정가 15,000원

ISBN 979–11–87244–93–6 (53410)